REWILDING THE URBAN SOUL

Claire Dunn is a writer, speaker, therapist, guide, barefoot explorer, and passionate advocate for rewilding our inner and outer landscapes. She worked for many years as a campaigner for the Wilderness Society and now facilitates nature-based personal development and leadership through rewilding, deep nature connection, and contemporary wilderness rites-of-passage. In 2010, Claire lived in the bush for a year as part of a wilderness survival program, an experience she wrote about in *My Year Without Matches*. She currently lives in Melbourne.

www.naturesapprentice.com.au

REWILDING THE URBAN SOUL

searching for the wild in the city

Claire Dunn

SCRIBE

Melbourne • London

Scribe Publications
2 John St, Clerkenwell, London, WC1N 2ES, United Kingdom
18–20 Edward St, Brunswick, Victoria 3056, Australia
3754 Pleasant Ave, Suite 100, Minneapolis, Minnesota 55409, USA

Published by Scribe 2021

Typeset in Bembo Book MT Std by the publishers

Printed and bound in the UK by CPI Group (UK) Ltd, Croydon CR0 4YY

Scribe is committed to the sustainable use of natural resources and the use of paper products made responsibly from those resources.

978 1 913348 72 4 (UK edition)
978 1 950354 78 8 (US edition)
978 1 925713 15 2 (Australian edition)
978 1 925938 93 7 (ebook)

Catalogue records for this book are available from the National Library of Australia and the British Library.

scribepublications.co.uk
scribepublications.com
scribepublications.com.au

This book is dedicated
to all those who champion the Wild

And to my loving parents,
Bob and Pauline Dunn.

'Wildness is a form of sophistication, because it carries within it true knowledge of our place in the world. It doesn't exclude civilisation but prowls through it, knowing when to attend to the needs of the committee and when to drink from a moonlit lake.'

Martin Shaw, *A Branch from the Lightning Tree*

Contents

Prologue
The Choice

'You're moving to the big smoke? Really, girl?' Mark looks at me incredulously, covers his mouth with his hand, and turns away in a half-mock theatrical giggle. It turns into a gravelly cough, which gives weight to his hanging question. I can't help but laugh too as he tries to regain his composure, prematurely looking back in my direction before being beset by coughs again. Finally, he turns towards me, his head cocked on one side with a look that is more of a listening, his dark eyes gently enquiring into mine.

Behind him, a small plume of smoke still issues from the welcoming-ceremony fire circle that Mark extinguished not long ago. Small groups of people stand sandalled or barefoot on the grass, chatting quietly. A few kids are chasing each other around the single pecan tree in the field. In the background, the forest is dark along the creek line, lightening to blues and greens as it meets the sandstone.

The ceremony was the most animated I've seen him — striding confidently through the space made by a hundred-odd people in a circle, gesticulating with his arms as he talked. His passion for the country here emanated through his every gesture and footfall. 'You've got to claim your belonging,' he said. 'We've all got to belong to this land, everyone, everywhere. Then we are the caretakers of old.' The crowd around him was captivated.

'An elder in training,' he used to joke in humble introduction. But it was a half joke. He's apprenticing still to his old uncles and aunties of the Gumbaynggirr clan, who reside on the north coast of New South Wales. I can feel some threshold has been crossed in the three years since I've seen him. It's in his gaze, like he's seeing through me with X-ray vision.

I feel a lump grow in my throat as I meet his eyes, remembering the first time we met. It was the first week of a yearlong wilderness survival skills residency that I was embarking on with five others on a piece of land not twenty kilometres away as the crow flies. I can clearly picture myself, walking into the thick of the smoke, directing it over my head with both hands as if splashing water, my heart a cauldron of fierce intentions for the four seasons ahead. And Mark there, tamping down the gum leaves onto the coals with his bare hands. He would return throughout the year at intervals, walking us out on country, showing us the plants that would feed and heal us, cooking up some damper when we got too thin, tempering the intensity with his belly laughs.

'Really, girl?' he asks again, this time more seriously. 'But I know you, you're made of the bush,' he says with a slight edge of concern. I'm amazed how well he remembers me after the passing of years.

Tears come to my eyes as I clock the truth of this in my body, the comforting smell of smoke in the air, the stirring cry of the channel-billed cuckoo, my hair still wet down my back from a dip in the creek and my bare feet spread like moulded clay on the earth. The majority of my years have been lived outside capital cities. The times I've lived urban, I perched rather than dwelt, a bird on a wire waiting for the wind to blow me back to a branch in the woods. And now Mark is witnessing me about to enter another kind of jungle, this time of the concrete kind.

'I know, it's kinda crazy,' I say vaguely, 'but I've gotta go.'

Mark looks on, waiting.

'Oh, okay, and I've met someone,' I say, smiling shyly. I don't tell him I've only known the guy for a few months.

'Ah, now we're getting the full story,' he says with a laugh, his eyes sparkling with mischief. It's not really the full story, I think to myself. There are other reasons, but they might be harder for Mark to understand, or at least to joke about.

'What's he gonna hunt you in the city?' Mark laughs and I join him, trying to picture my new man catching anything other than the mosquitoes he chases down with vehemence in the bedroom. Mark's wide smile fades and his face turns suddenly serious.

'You stay wild, eh? You're a woman of the earth. Don't forget what you've learnt out here.'

My full trolley is a reluctant cargo ship in the swell of a busy IKEA Sunday. Another laden trolley capsizes in my path. I stop to wait while the giggling bunch of uni kids attempt again to pile on a bedroom's worth of cheap furniture and accessories in the most counterintuitive order. Their trolley bumps mine, and the large flat-packed box at the bottom of my cargo slides onto the shiny white floor. 'Crap,' I mutter, and I try to wedge it back on.

I look around and try to remember which direction I've come from. I identify five possibilities, all looking equally unfamiliar. I squint under the overhead lights. No sun to guide my way here. Another load, captained by three more students, makes pace towards me. I've got to steer my ship out of here.

My trolley makes a resistant squeak as I turn and find myself in domestic dollhouse avenue — kitchen after kitchen curated to particular eras and genres. I pass farmhouse chic, ode to stainless steel, 1950s kitsch, contemporary warm. I half expect to see Ken and Barbie tossing salads and waving at me. This is not just buying kitchen furniture, or even an aesthetic, it's buying a dream, an off-the-shelf life.

Oh gawd, get me out of here. I've been wondering around for what seems like hours, but I wouldn't know, as there are no clocks, nor any clear up or down, just endless spirals. I feel like I'm stuck in an Escher painting. I'm thirsty and my heart is beating faster than the physical exertion of trolley hauling requires.

I had no idea what I was getting myself into. A friend suggested that I come to IKEA to buy a cube shelf to fill an awkward space below a window that was proving hard to fit with any second-hand furniture I could find online. This is no shop, though, more of a modern cathedral for the religion of Home. But here I am, another worshipper amongst the thousands. I look down at my trolley. The one piece of furniture I came here for is down the bottom, under a pile of other stuff — a full-length mirror, a small chest of pine drawers, sheets, a soap holder, a toilet brush. Stuff.

I eye the four custom-made cane baskets that slot into the cube shelves. Whose fingers have woven these? Which plants have these been harvested from? Who loaded them onto trucks or trains or ships? And where? There is no explanation on the tag. They have appeared out of the magic Santa sack of IKEA.

How did I get here? This isn't the life I envisioned when I moved to Melbourne two years ago. And what I'm shopping for is a very different nest to the one I was feathering seven years ago. My homemade shelter is so vivid in my memory I can almost reach out and touch it. The saplings bent over to form a neat dome, lashed together with the rough string I twined in the early morning when the acacia sap flowed supple and wet. I can hear the particular kind of rustle the blady grass roof made when its thatched braids were tussled by the wind. Sometimes I'd lie in bed during a winter storm and wonder if it was going to all blow away, leaving me exposed to the steely stars. The fire is bubbling and squeaking in cosy ambience, and I feed it slowly with sticks from where I sit on my grass mat, drinking tea. Seaweed hangs from the rafters, drying.

Looking around, I can see the home furnishings and trimmings that I designed and made with my own hands. Within them are the stories of the days at dawn harvesting vines to weave baskets that would become my wardrobe, pantry, and foraging bag; the quiet contentment of sitting around a fire blowing coals onto wood to burn out a bowl; the three days I sat unmoving on the ground and carved spoons and spatulas from a fallen branch; the sculpting of cooking pots and bowls from clay. I remember well the day I spent laying rocks in the centre of my shelter for the fireplace, cementing them with clay from the creek. I recall the imagination I employed to craft a wooden hook and pulley system for my billy to hang over the fire.

Out was symmetry, corners, and rulers. In was makeshift, inventive, and homespun. It amazed me how little I needed to live well, and to be comfortable. One of my fears was of not having soft furnishings. It was the last thing I missed, and the first thing I didn't. A thickly woven grass mat and a backrest felt positively luxurious after an active day outside.

It was an in-joke amongst us on the residency to compare our 'primitive IKEA ware' — having never stepped foot in an IKEA store before but associating it with a kind of simple wooden aesthetic. But here on the shiny showroom floor, steering a trolley full of stuff that could have come from Siberia, I feel about as far from that world of my creation as I can imagine. If I'd had a crystal ball back then, I might have thrown it in the billabong.

I pass by a wall of cushions and pillows. They threaten to suffocate me in their downy wads of feathers and filling, a symbol of our collective somnolent addiction to comfort. My stomach grows queasy. I search for the nearest exit sign and head directly for it.

Back home, I read the label on the chest of drawers and realise that formaldehyde was used to stick the particle board together.

That means I'm probably breathing it in right now. Still, the low cube shelf fits perfectly, which is not easy in this odd-shaped room.

My new flatmate Min bounds down the stairs to see how I'm going.

'That was a nightmare,' I say.

'Oh yeah, IKEA is its own form of beast,' Min laughs. 'But that looks like a good fit. You've created a beautiful wild little cave down here. The basement has been transformed!'

'Getting there,' I say. Last weekend, I ripped up the grotty carpet, sanded and polished the floorboards, and painted the walls white. I converted the strange little alcove into a bed base that extends out into the room. Sitting up in bed is like resting on a ledge at the back of a cave and looking out the overhang into the forest.

'This must be your book,' Min says. I glance up to the see the all too familiar spine of *My Year Without Matches* being coaxed from the bookshelf.

'That's the one,' I say, feeling a little shy.

Min flicks to the photo section in the middle.

'Wow, your shelter is so beautiful. I would have loved to have known you in this time,' she says, angling the book in my direction to show me the image of my head poking out of the top of the tiny remaining hole in the thatched roof. I'm wearing a big hat and an even bigger smile.

Min glances up and catches my eye for a second. 'I can't imagine what it's been like to move back to the city. To be so grounded in a place, to be in connection so completely, and then ... well, the city is a different type of wild.'

I nod, not quite knowing what to say.

'Can I borrow it?' she says, holding up the book.

'Sure,' I say. The book and Min disappear back up the stairs.

I pull some items out of a box and start arranging a loose mandala of bones, stones, feathers, and shells on top of the IKEA shelving. Inside the box, I make contact with something rough and cylindrical — the banksia cone I collected from my shelter site. The only physical item I brought back with me.

Something about this particular house move is stirring the whole pot of that chapter.

If I were to answer Min's implicit question, I would probably start with the most explicable and tangible layer first — the story of the burnt-out environmental activist who took off to the bush to heal and reconnect with her original love. It's partly true. After almost a decade of dedicated grassroots campaigning for forest and climate protection, I was becoming a 'greenocrat', spouting statistical certainty from my office swivel seat in the city of Newcastle rather than from the immediacy of a forest protest.

There's only so long that kind of fire can burn. The flames cooled to embers. And from the coals a new burning question grew. *What's the real reason we're losing this battle?* I asked myself. *What really creates change?* The strategy thus far — shout the truth from the rafters until people hear it and join you — had proved limited in its efficacy. It seemed to me that change was less about a lack of information, and more a lack of love. Disconnection from the natural world makes it easier to treat it like a commodity. Trees are things we can rip out and use, without realising we're taking from our larger body too. Facts don't move people to act. Love does.

I wanted to fall in love again.

I travelled to the US for two summers to work and study with famous tracker and survival expert Tom Brown Jr, himself once an apprentice to a Native American elder and master tracker. I sat in traditional sweat lodges with songs sung in first languages. I sat my first Vision Quest, a four-day solo fast on a mountain. I spent seventeen days camped on the edge of a national park on my own. I tested my skills on multi-day survival trips. I travelled to Arnhem

Land and sat with the women on land that had sustained an unbroken lineage of their ancestors for countless thousands of years.

Something in me was waking up. I grew more curious. What might happen if I stepped in fully? If I was able to *really* steep myself in nature? Immerse myself in this language older than words; enter into this largest conversation one might have with the world? One that's far from to-do lists, strategies, or small talk, but powered by the immediacy of bird, lizard, rock, sky, dewdrop, imagination, snake, dream, rain, spider, song, mountain, and stream? What would happen if I uprooted myself from the city's conveniences, comforts, and distractions and plugged myself back into the primary relationship of earth and self? There was some treasure, I knew, that could only be found when my imagination was fed not by the same social mirrors of my current life but by the untamed and unshod. I knew I had to enter into a deeper relationship with wild nature to find out. But how?

I was invited to take part in Australia's first Independent Wilderness Studies Program, a loosely organised yearlong residency where a small group would apprentice themselves to the arts of earth-based living. There were to be few visitors in and few trips out. There would be some structured learning, but mostly the schooling would be in unschooling — running feral and free. It was as much about encountering nature as it was about encountering oneself in nature. Unlearning is what I needed: to unpick some of my civilised conditioning, step beyond the four walls of certainty, and experience life unmediated, unwrapped. To feel how fire feeds and warms me not by the flick of a switch but by the blisters on my palms from drilling wood onto wood to create a spark; to taste the earth in fruit, shoot, and berry; to track the seasonal changes; and to rest with the dream-rich depth of a nervous system not peaking on caffeine, sugar, stress, and overstimulation.

I was the first to sign up. Not many months later, I found myself clearing a space to live under the limbs of a giant grandmother banksia tree, the seed pod of which I now hold in my hand. I make another space, this time atop the new IKEA shelf, and carefully lower the cone.

I think of Min upstairs in her loft turning the first few pages. That the story of my journey that year can be found in a bookshop still amazes me. It was so hard to try and capture in words, so full of paradox and dichotomy — both extraordinary and yet very ordinary, practical and mystical, ecstatic and struggle-filled.

A white-browned scrub wren calls outside close by. I wonder if it's raising the alarm over the neighbour's cat; I pop my head above window height to see the tabby's tail disappearing under the privet. Caught! It feels good to be able to know these things. I've already clocked the fox prints down at the river, and the perch the spinebill likes to sing from near the letterbox in the morning sun.

As I had hoped, my year without matches did indeed gift me entry into this larger conversation with nature. Sitting and watching attentively day after day, I became porous and receptive to the patterns and routines of the forest. My awareness busted out of its accustomed ruts, my default shifting to a full-bodied sensory engagement. To some extent, I'm still listening from this place. And can switch it on when I want. The abiding curiosity that turned on out there is as strong as ever, the book of nature a never-ending fascination.

But love's promise is as much about descent as it is ecstatic union. This was the part of 'change' that I hadn't bargained for. The forest cracked me open, a hard seed pod that fell to the ground and over the months and seasons was worked on by the slow action of wind and rain and soil microbes.

All my youthful certainties, all my conditioned habits of achievement and perfection — the entire house my ego had carefully constructed so as to experience the world with as little vulnerability as possible — dissolved. It wasn't just unlearning, it was unravelling. I was stripped back to bone. The long wanders in the forest where I would lose myself were a mirror to the inner disorientation. I didn't know who or what I was anymore. In that painful undoing, what emerged was a less cultivated, less managed self. My compass bearing shifted away from the directionality of expectations, judgements, and shoulds and towards intuition and instinct. My capacity to feel *everything* expanded.

The irony is that the trail I followed to the bush has led me back to the city. A few years after the residency, my curiosity turned towards my own species again — to know what mysteries lay behind those uniquely human eyes. I'd been a lone wolf long enough. I was ready again to swim in a sea of humanity, to let myself be enchanted by, and in turn enchant, another human heart. I was ready to feel my gaze returned with a similar understanding, to know and be known in the way particular to my species. I was ready to find edges and challenges of a different kind.

The next wild adventure beckoned, drawing me into the centre of human cultural flowering. I knew in this I would also be saying yes to concrete rather than dirt underfoot, streetlights rather than moonlight, freeway noise rather than cricket song, busyness rather than idle hours.

I said yes because I had made a promise at the end of my bush year. It had come to me during one of my last overnight wanders in late spring, when the ground had warmed enough for me to sleep without padding and the sun shone strong enough to warm my bones at dawn. Drunk on awe and gratitude as I watched the sunrise, I realised I needed to build a bridge, to bring what I had found out in the bush back to the city, to ignite the spark of the wild within the heart of a disconnected culture.

I just needed a carrot to get me there. It came in the form of a man.

Yet after a couple of years, I was still unsettled, both in place and person. I went bush a lot, running or attending workshops, a growing point of contention. The online demands of his startup business were at odds with my outdoor pursuits. I wanted to put roots down but felt like there was no real soil for them. Soon, the man and I parted ways.

Ostii and I had met in the backyard of his inner-city townhouse under a wisteria dripping with sweet bunches of purple flowers buzzing with bees. I was in Melbourne promoting *My Year Without Matches*. 'That colour suits you,' was the first thing he said when he saw me in a mustard jumper, my wet hair twisted up in a top knot. I took in his prayer beads, zip-heavy canvas vest, scruffy brown hair, and piercing blue eyes. The gaze that I returned was as unwavering as his. Three months later, he flew north to help pack my belongings into my car and then serenaded me the thousand kilometres to his tiny city castle.

As we left the flat plains of the interior, the city grew like a grey bubble on the horizon, the jagged skyline of the skyscrapers slowly taking form as we sped towards it with all my worldly possessions. We both grew quieter as the suburbs knitted together. It wasn't until we pulled up outside his home, my new home, that we looked at each other and burst out laughing. The wild ride was just beginning.

We found a rhythm together that was threaded with reams of laughter. We planted loquats and mulberries on the nature strip, filled the backyard with tomatoes, rocket, and basil. We danced and wrestled in the living room and entertained around a tiny fire pit. In the mornings, I would take my cup of tea and squat in the stamp-sized patch of sunlight in the backyard, watching the white-naped honeyeaters suck nectar from the single gum tree. Yes, there was a lot of good in those years.

I contemplated leaving the city after we split but realised I wanted to remain. While the countryside is rich in natural beauty, I had struggled to find a place for myself in the towns and villages of regional Australia. I liked playing in a bigger pool, alongside the 86 per cent of other Australians who call the city home.

When Ostii and I had driven into the city that first time, and I looked out the window past the countless anonymous faces buzzing about on their business, I wondered what draws so many of us here. Surely not the waiting a year for a parking permit, waiting in lines at check-outs and cafes, waiting in peak-hour traffic. Noise. Pollution. Limited green space. The rushing and the busyness. Not even the good coffee or the availability of sushi at 3.00 am.

I imagined the city like a giant beehive. If the cooperative nature of the bee colony produces honey, so too a city-hive with its density of diversity — of people and cultures, organisations, resources, and needs — is an incubator for creative nectar. Cities drip with innovation and invention, the sheer weight of numbers offering endless possibilities for collaboration within more defined niches, micro-environments of specificity where synergistic possibilities expand exponentially. Beehives contain a kind of pressurised communality that happens to yield one of the sweetest products on the planet. So too cities provide a certain creative tension, a cooperative-competitive edge when so many people rub up against each other.

I suspected that there was an aspect of wildness I had yet to explore — the human creative kind. I was ready to play in the particular kind of playground that only cities provide, to dance in the anonymity, and experiment in the laboratories of cultural diversity and depth.

Now I am one of the bees in the city hive. Ostii is gone, but I have found a house to share with Min and two other creative busy bees. Seven years ago I didn't own a single key; now my keyring jingles with a dozen. The $2,000 I survived on for the bush year wouldn't get me much past a month. Gone is my phoneless freedom; my iPhone has more apps for instant communication than I have fingers. Whereas once my job description was the 'sacred order of survival', now it includes issuing invoices and paying taxes. My daily errands no longer include water collection, basket weaving, and eel trapping. They spill over the page and include engagements with dozens of colleagues, clients, collaborators, and contractors. And my nature time has shrunk from all the time to never enough.

Yet if I'm honest, there's a part of me that clings to the identity of the bush me as the real me, that thinks this city life is some sort of phase or experiment. I'm a city dweller with a wild heart and I still haven't found a way for these aspects to happily coexist. And I need to, try at least. Because if I can't, then how can I be that bridge? How can I expect us to fall in love with the world in the way that's so needed if it's dependant on going bush for a year? No, it has to be possible, right where we are. How, I'm not exactly sure. But it starts with asking the question. And I'm not the only one with this thought on my mind.

I catch a National Geographic documentary titled *The Call of the Wild*. It follows nature writer David Gessner as he pokes his nose into the lives of those exploring the current state of the human–nature connection. After brainwave mapping the interference of phones during a walk in the park, and a spot of forest bathing, Gessner gets serious, joining a guy named Colbert Sturgeon in his handmade campout on the banks of a swamp in Georgia. Once an insurance middleman, Colbert left it all to live off the land, dropping virtually all contact with the monetary economy. Tucking into a meal of barbecued snake with Colbert on the deck of his riverside shelter, Gessner reflects on how we are at a historical tipping point. More than half the world's population now lives in urban areas, a proportion expected to increase to 66 per cent by 2050.

'We're in a petri dish, undertaking a huge experiment,' Gessner says. 'What happens when we remove ourselves so profoundly from nature?' The assumption here being that the evolutionary blink of an eye in which we have moved off the savannahs and into townhouses has not dampened either our biological or spiritual suitability to live intimately with our environment. Rather than disappearing, the capacities have atrophied, become lazy, been forgotten about.

One possibility within this petri dish, says Gessner, is that we exist increasingly within a bubble, our ancestral lifeways becoming more and more distant, until such time as the bubble bursts. The title of the documentary, however, points to the more optimistic direction Gessner is heading: the 'call of the wild' is hardwired in us, and the degree to which we have strayed from our evolutionary origins is triggering an equal and opposite reaction — in the direction of nature.

It's a movement some are calling 'rewilding': a reclaiming of a more wild or original state, a movement away from domestication. The word itself speaks a thousand, its simple conjunction of 're' and 'wild' pointing to the fact that we were all once wild and can be again.

The word has its origins in conservation biology, referring to the strategy of reintroducing keystone species such as apex predators. The most famous example is Yellowstone National Park, where the reintroduction of wolves caused a trophic cascade of beneficial ecological effects that extended down even to the river invertebrates. The absence of wolves had caused a boom in deer populations, who then ate out the valleys. The willows had been reduced, and subsequently so had the beavers. As deer numbers fell under predation by the wolves, beavers returned, and their activities shifted the stream hydrology. And that was just one chain of events. There were many. Life came back into balance not through isolating nature behind fences but through upping its wildness factor.

Just as this rewilding movement seeks to restore the agency of ecosystems, the human rewilding movement is seeking to restore balance and vitality to our inner ecosystems. For as we have systematically controlled and tamed nature, we have simultaneously controlled and dewilded ourselves. The result is a range of physical and psychological ailments that are hallmarks of modern life — rising levels of suicide, depression, obesity, and disease; childhood behavioural problems going by such labels as ADHD; the loneliness epidemic that underpins a collective spiritual crisis of meaninglessness. Our once wild, bushy tails are now threadbare and dragging behind us.

Rewilding writer and practitioner Arthur Haines draws a parallel between captive animals and modern humans. Though lions held in a zoo have greater life expectancies and longer life spans, he says, most people understand that such a lion does not live the life it was biologically intended to. While it is still a lion, some of its 'lion-ness' has been taken away. It's precisely because of the hardships and uncertainties of life in the wild that the animal fully lives its wholeness. So too, Haines says, the chronic disease and emptiness of modern culture indicates we are not living as humans were biologically intended to — some of our 'human-ness' has been taken away. Caged by convenience, we modern humans have forgotten our wild roots, our acute capacities of perception, our kinship with the natural world. We're asleep on the doormat while the world burns.

Rewilding asks us to consider afresh the link between wildness and health. Both Yellowstone and my own experience tell me that given some encouragement, the wild will return, that it even longs to return, waiting with poised claw and sharpened beak to step back into the great dance with us. Rather than chaotic or dangerous, wildness in this context is a healer, an agent of balance, a wisdom keeper in its own right.

What if wildness itself is the overarching keystone quality that could revivify our modern human ecosystems?

Perhaps the real endangered species is that of the wild human. Winning that campaign would negate the need for all others. For this species, if reintroduced, could return health, vitality, diversity, and abundance to our cultural and physical landscapes, to the centres of civilisation themselves.

Unlike the back-to-the-land movement of the 1970s, rewilding is as much urban as it is rural — is more about the cultivation of wildness as a quality and experience than about turning back the clock to some perceived hunter-gather idyll. Rewilding asks us to question our allegiance to domestication, to reconsider everything from our food to our footwear in light of our evolutionary lineages, to make brave gestures to court the wild back into our lives and hearts.

'We've got to find a way to bring the wild back, into our cities and into our lives. Even small doses of the wild can have enormous benefits,' says Gessner. 'Now more than ever, we have to heed the call of the wild. And if you look closely, you can find wildness anywhere, even in your own backyard.'

I had chosen to be in the petri dish with the five million others I share this city with. What might it look like to reclaim elements of thriving from our biological and cultural heritage that we've lost in this urbanised world? What is possible to reclaim of this ancestry within the limits of the urban matrix?

The city is nature too, its wildness found in different forms. I want to unveil the hidden connections, make visible the lineages to the elements, creatures, and stories that form part of my life here. It's time to fly down from my perch and sink my talons into this city, to give up peering over to greener pastures, to renounce the rural dreaming, the nostalgia for the shelter now rotting on the forest floor.

I'll open my eyes and ears to the patterns and mysteries of nature where I live, make links between my life-support systems and their origins. I'll experiment with ways to experience myself not as a separate entity but nested within communities — both human and those of the larger biota. I'll ask myself the question of what connects me and what disconnects me and move towards greater connection. I'll clear a space in this concrete jungle and see if this wild body, wild heart, and wild mind can make a real home here.

This is my next challenge, a challenge perhaps harder than my year without matches but equally important. To learn to answer the call of the wild right here, where I am, where we are. To find the wilds within myself and within the concrete labyrinth of city streets. To rewild the urban soul.

In the end it's a choice. A choice to belong. A choice to pay attention. A choice to love.

Chapter 1

'A little further, couple of inches along, almost there,' Phillip says
with his face turned parallel to the ground. I heave myself against
a wombat-sized rock. It groans and obliges by rolling over on its
side, shouldering in close to its neighbour. That makes six. Shiny
from the morning dew, the stones could be a row of smiling black
teeth, pockmarked and chipped, from a giant long dead. Behind
them sits the tongue — a perfect circle of freshly levelled earth
that Phillip and I have been creating over the day. It too is dark
and muddy, the grassy slope excavated by our shovelling hands in
service of a wide and flat fire circle to gather on in the backyard.

Our surveying consisted of a piece of string that marked out
the area that we dug into. The rocks provide downhill retainer.
This morning, they woke up at the top of the hill around the
garden at the back of the house. To their surprise, and hopefully
delight, they were hand-picked on merits of roundness and size
and sent tumbling one by one downwards down to their destiny,
albeit with much coaxing and coaching. Instead of gathering moss,
my rocks had a penchant for gathering blackberry canes, and many
a time I found myself up to my arms in the blackberry thicket
trying simultaneously to keep myself from being swallowed by
the thorny monster while foot-rolling the rocks into the clearing.
Phillip's rocks, however, seemed to find much smoother passage
around the side. Despite the cuts and scratches, rock rolling has
been a remarkably satisfying way to spend a Monday.

'I feel like a dung beetle,' I say, stretching my arms up over my head.

Phillip laughs. 'Rollin', rollin', rollin', keep those dung balls rollin'.'

He stands up, so tall his shaved head is almost tickled by the lowest branch of the red gum. His single silver hoop earring catches the midday light and glints.

'It's looking great,' I say as I join him in admiration of our hard work.

Phillip's a friend who has moved into the house on a nine-month sublet, joining me, Min, and the longest-standing household member, Megan. His hands speak much about his particular genius. With a gap between carpentry projects, I proposed a regular work n' play date. We made it Mondays. It's a great way to start the week: when everyone else heads back inside, we're like kids creating our own version of an Ewok village, making gardens, planting trees, rolling rocks, messing about. There's no shortage of plans, the two of us full of big ideas that only get bigger when in each other's company. It's helping me settle in too, all this ferreting around the pockets and crevices of the yard, this making and shaping of my habitat.

The first project was upping our edible garden capacity in the front yard. Phillip is a master urban gleaner of all things building-related, with a foxy sixth sense for discarded treasures. Hard-rubbish throw-outs are his Bunnings. When he gets a hunch about something, he'll stand back with hands on hips and bite his bottom lip until the idea is fully formed, then launch into action. Last week, the hunch was about garden sleepers. 'Come on,' he suddenly announced, and drove us a kilometre up the road to his old share house, now undergoing renovation. We slipped in through a gap in the scaffolding and waded through waist-high grass down the side path to the backyard. In the shed, a dead bird was mobbed by flies. I brushed against grey foliage and was struck by a bitter scent. 'Here, put it under your pillow tonight,' he said, breaking me off a sprig. 'Mugwort — brings on dreams.' Clearing away the grass in the centre of the backyard, Phillip exposed the remains of his old fireplace, where he used to gather friends for equinox and solstice ceremonies. (As well as building, sharing seasonal celebrations is something we love. The reason the new fire circle is so large is for this purpose.) At the back of his old yard, we dug up the remains of red-gum sleepers. Some were too far gone, but we repurposed four for the new garden beds.

'Break time?' Phillip asks, gesturing with a jerk of his head in the direction of the river. I nod. He takes my hand, his calluses rough against my skin. We wander through the neighbour's yard and down some steps onto a jetty that juts out into a wide brown river.

The Yarra, Melbourne's famous watercourse, right at my back door. It's already travelled a ways to make it here, starting from a trickle in the mountains, passing through the valley hinterlands and the eastern suburbs. It holds strongly to its wild origins here, both in depth and velocity, but not too many twists and turns away, it meets a creek that has collected the city's northern refuse and a weir that tames it significantly. To my right, the river curves sharply to the south. Upstream, the straight stretch of river is framed solely by the reflection of river red gums. It reminds me of one of the bucolic Australian bush paintings hanging in the hallway at the family farm. The bank opposite is bush, part of the parkland that hugs the river for kilometres. I drop to my knees and dangle my hands in the water. The cool eddy swirls through my fingers.

I still can't believe I live here. And that this is only five kilometres from the CBD. Waking up most mornings since I moved in, I have had to remind myself that my backyard includes a river. And in fact, there is *nothing* human between my bedroom and a wide expanse of water, not even a fence. In the mornings, I've taken to rolling out of bed, out the back door, and down the switchbacks to the river flat, where I can squat close enough to the water's edge to breathe deeply of its scent, let it touch my skin, and give gratitude for all the waters of the world.

When I return to street level, knowledge of the proximity of this unimpeded place of nature stays with me, carving out a corresponding internal spaciousness. I continue to roam and inhabit the riverbank long after I've left, its imprint strong enough to linger, aided by the occasional loud shriek of the swamp hens or Pacific black ducks that drift up as a reminder.

It was a synchronistic chain of events that led me here. I wasn't even looking for a new place at the time. It was more like the place came looking for me. I had met Min through a mutual friend and she invited me to come for a kayak with her. I didn't imagine she was meaning in her backyard. I was familiar with the suburb as the fringe edge of the trendy inner north, increasingly populated with creative and hipster types heading east in search of cheaper rents and greener pastures. I hung a right into the street from a noisy arterial road and parked outside the wisteria-covered picket fence. Entering the front gate was like entering into a kind of oasis, one that became more like fantasy when Min guided me down the hill to the shores of Melbourne's famous Yarra River.

'What, you can live here? I thought this was just for the rich and famous?' I said in disbelief.

'Yeah, and us,' she laughed. 'We call it the Riverhouse.'

The old three-storey weatherboard dwelling was light and airy, and tied together with string and sticky tape. It had been a share house for as long as anyone could remember and was still signing away leases for less than you'd pay for a bedsit in the CBD. I received a text from Min a couple of weeks later. The longest-remaining housemate had announced he was moving out and would I like an interview. I fell back on my bed, rested the phone on my chest, and felt the ground shift under me. I recognised this kind of moment, when life suddenly forks and a new path pulsates with vitality. To refuse would be to swim against the flow of the river. To say yes, though, meant leaving the other branch behind. There was loss at the juncture. And the inevitability of change.

'The Birrarung,' Phillip says, with his hands on his hips, referring to the traditional Woiwurrung name of the river for the Wurundjeri people, the original custodians.

'River of mists and shadows,' I say, completing the translation.

A city's personality is defined by nature and culture and the way the two intersect. When I lived in Sydney, waves of us fell upon the stunning coastline and harbour, a purported mirror to the sunny, somewhat-flashy, and extroverted disposition of the place. This river city with its wetter, colder climate lends itself to a slower, more egalitarian pace and a blossoming of inside artistic pursuits.

The current name for the river has its roots in the Wurundjeri term 'Yarro Yarro' from the Boonwurrung language. Meaning 'ever flowing', the original name was turned into Yarra Yarra by the colonists and then simply the Yarra. The lower stretches of the river beckoned the early settlers in 1835 and thus began the displacement of the Indigenous people from their homelands of at least thirty thousand years. From its source in the Yarra Ranges, the river flows 242 kilometres west through the Yarra Valley before opening onto the plains of greater Melbourne and eventually the bay.

While I'm one of the few who get to live on its misty banks, its presence and that of its many tributaries are near and dear to this city's inhabitants, who gravitate towards it to walk, cycle, picnic, and boat.

Kek Kek Kek Kek. There's that bird call again! What is it? The last few days, my ears have picked up on the same shrill, staccato call that I assume must be a raptor of some kind. I've run outside to try and spot it but haven't yet caught sight of it. *Kek Kek Kek Kek*. There it is again, closer by. I jerk my head back to catch the distinctly pointed wingspan of a bird of prey flying directly over us.

'Look!' I say, pointing. Phillip follows my finger. Without even a flap of its wings, the grey bird glides silent as an owl into the top branches of the gums a few blocks up, disappearing from view.

'Awww, cool!' Phillip says. 'Did you see its wing tips? Looked hawky to me.'

'Definitely a raptor!' I exclaim.

Could there be a bird of prey nesting nearby? That would be amazing. I'm going to need to keep my eyes and ears peeled. What else might I find in this pocket of wildness?

Early afternoon bleeds into late. The sweat from the day's work is sticky on my skin. I glance down at the river. I haven't been game enough yet, but this could be the moment.

'Swim?' I say. Phillip stops work and raises his eyebrows. It's a mild response compared to mine when my friend Elizabeth first suggested a dip. 'In the Yarra? You've got to be joking,' was what I said.

Early industry, including a now-closed paper mill just upstream, contributed significant pollutants to the river in the way of chemicals, grease, oil, and heavy metals. For the last few decades, though, industry has relocated away from the river and now the major pollutants are stormwater run-off, sewage — and whatever remains of the previous pollution. Was it clean enough for full-body immersion? I was dubious.

Elizabeth directed me to the Yarra Watch website, which publishes weekly water-quality observations specifically for recreational swimming. Testing is performed at four sites along the river during the summer months, including at an 'urban' testing site a few hundred metres upstream from us. Apart from the few days after a rain event, the website seems to give a consistent 'fair' rating. If the government is willing to risk a thumbs up for a suburban swim, then it can't be too bad. Phillip doesn't need much convincing: his risk threshold is even lower than mine.

'Let's do it!' he answers with a huge grin.

Still, I'm a little squeamish as I strip to my underwear on the jetty and watch a plastic water bottle and muddy tennis ball float past. I try to blank out the images of dog poo from my mind and inch my toes closer to the edge.

Phillip takes his hearing aids out, places them neatly on his clothes, and slips into the water to check for snags. Bubbles rise to the surface as he dives below, emerging with a shake of his hairless head. He gives me a thumbs up. I peer below into the brown swirling mass, wondering where it's come from and what it might have collected along the way. Can I really trust this water? Not so many generations ago, you could drink from it. But city rivers have gained a reputation for basically being drainpipes of the suburbs. How could this be any different? Am I really about to swim in the stormwater drain of every suburb that passes through the miles and miles of both the eastern and north-eastern suburbs? A Melbourne-born friend told me he came out in a rash after swimming here twenty years ago. What's the shelf life of some of these pollutants?

'Come on!' Phillip calls from where he floats in the centre.

The allure of a backyard swim wins out. With a squeal, I launch from the jetty and descend in a giant dive bomb. My legs begin to kick, and I pull my undies back up as I rise. I surface with a yelp of shock and exhilaration and blow the water droplets from my lips before they dribble inside.

Phillip is grinning like a pirate.

A kayaker glides past. 'You're game!' he says.

'Yep,' I puff as I catch my breath, feeling a mix of reckless and a little foolish. A duck appears and cruises past within arm's length, looking completely undeterred by my presence. Perhaps my willingness to join it in the muck and tumble mitigates its wariness of the human form. I swim around in a 360. Wow, the river from duck height is entirely different. It's like I'm seeing it in panorama, the surface ripples now an embodied rising and falling of waves, the trees protectively paternal in their overhang.

'See if I can find bottom,' Phillip calls before disappearing underwater. He reappears with a shrug. 'Nah.' It's as I thought, though I can't see it, sensing a depth beneath my dog-paddling feet.

I gather some more courage and breaststroke upstream, every underwater face-full without incident building my confidence. A Eurasian coot swims right up to me, like the duck did, without any timidity, as if the act of being in the water has removed my predatorial humanness. 'Hello, cooty coot!' I call. I start laughing, I'm swimming in my backyard! In the city! It seemed like such an urban myth. Unless I come down with a rash or the runs, it's a myth I'm willing to promulgate too.

We swim over to the other side and clamber up onto a fallen log.

A grey shrike-thrush flies in, alights the branch above, and calls loud and strong. Even Phillip hears and smiles. The first week here, I got a bit confused at the melodious and consistent bell-like chiming I could hear from the river until I realised it was shrikey calling in a dialect different to the one I was used to up north. 'Oh yeah, it's their city slang,' said my birdo friend when I asked him. I copycat whistle to the one on the branch. Shrikey tips its head as if considering and then issues a reply. Much to Phillip's amusement, our back and forth echo goes on for a few minutes before the bird tires of its language lesson and flies off.

Phillip is lanky and graceful as a stork as he strides along the log and up the bank towards the cherry-ballart tree. Its trunk is so brown it's almost black, and its long lettuce-green leaves stand out against the grey-green of the eucalypts. Ducking in under the low-hanging branches, Phillip emerges with a few sprigs of foliage, which he holds above his head as he swims back across the river. I follow in his wake. Dripping wet, we squat around the new ring of rocks on our fire circle as I hold out a bunch of dead leaves while Phillip lights them from underneath. The immediate warmth meets my skin with delight, and I shift a little closer to the flames. I am warmed from the inside out, not just by the fire, but by the day.

I'm reminded of a quote attributed to Jean-Paul Sartre that I once had laminated and stuck on my wall. Four neat numbered points laid out the 'Four Ingredients for Happiness' — 'Life in the open air', 'Love for another', 'Freedom from ambition', 'Creativity'. All four qualities are well represented in these Mondays, reflected in my sustained experience of joy and pleasure. I've extended the experiment by giving myself full permission to prioritise these four ingredients as much as I choose for a certain period of time. So far, it's been three weeks of mostly playtime outside and not much work. It's part of a strategy to use the house move to set a new normal. I'm determined to not get so busy that I miss the moments of magic, like the one with shrikey, that come only through a sense of space and playfulness. We'll see.

'The first fire,' I say.

'Yep,' Phillip says, with a glance that tells me he's been anticipating this moment. 'Our welcome ceremony, in fact.' I can tell he's waiting for some coals to form, and I settle cross-legged, watching as the sticks crackle to flame. Ah, fire, so so good. Everytime. Without fail. I think back to Mark's plea for me to stay wild. Fire is my magic pill for that, instantly dropping me into a connective place. It makes sense. Fire is perhaps the most consistent and universal practice of our shared human ancestry. Over 800,000 years ago, our early nomadic ancestors in the Ice Age would have gathered around flames every evening, illuminating the faces of those sharing stories and cooking food from the hunt. Until the advent of agriculture, everything centred around a wood fire. It's in our bones, the feeling of safety and belonging when around fire. Testament to this is the fact that it remains legal to have a small backyard fire even in the city, some recognition of the ongoing importance of access to natural flame. My friend Arian recently sent me a photo of the fire he has been cooking on over summer in his tiny London backyard.

Scraping out some coals, Phillip places onto them the cherry-ballart leaves. It's one of the plants we both know was used for a traditional smoke cleansing by the Wurundjeri people, particularly for welcoming. The green leaves begin to smoulder immediately, and I move in towards the billowing smoke, directing it over me with a scoop of my hands just as I did with the river water minutes ago. The scent is earthy and pungent. My lungs fill with it, and my out-breath is smoke-infused. I turn around and let it waft over my back in clouds.

The river here is thick with story and myth, I can feel it. Stories it holds for itself, that are the river's alone, and layers of meaning built up over the tens of thousands of years of human habitation. Right now, with the smell of the ballart, it's not hard to imagine a presence right here at this generous flat on the bend. When Phillip and I were unearthing the layers of soil for the fire circle, we imagined together the feet that have walked softly over this turf, the campfires that have been lit where we now sit, mussels and fish cooking on the coals. Could this have been a regular gathering place? A ceremony ground? What crafts and ancient technologies might this place have witnessed or inspired? What depth of ecological knowledge was once held here? How much has been lost?

Only a week ago, I sat on the banks of the same river surrounded by a couple of hundred other people for the traditional Indigenous Tanderrum ceremony. A patch of concrete had been reclaimed from a city park and covered in sand as a gathering ground for the Wurundjeri/Woiwurrung, Boonwurrung, Taungurung, Wadawurrung, and Dja Dja Wurrung language groups of the Eastern Kulin Nation. I learnt that the 'wurrung' on the end of each group's name means 'mouth' or 'lips', indicating that each group's identity is based on its language. 'Wurundjeri' is an exception to this rule, as Wurundjeri was the last remaining clan group of Woiwurrung speakers, and they now identify using this name. The original clans were Wurundjeri-balluk, Balluk-willam, Marin-balluk, Kurungjang-balluk, and Wurundjeri-willam. In times of plenty, large gatherings known as Tanderrum occurred between different language groups for the purposes of trade, initiation, marriage exchange, political discussion, and celebration. We bedded down on our jackets on top of concrete for front-row seats. Even though part of the city's annual arts festival, this was no performance. The songs and dances carried a potency and freshness that surprised me. The painted bodies morphed into waterbird, emu, and wallaby. Bunjil, the eagle creator spirit, soared. I watched the dancers waiting in the shadows as much as those in the spotlight. Their ochre-painted faces were enlivened, their limbs still twitching creaturely. They looked on, the women almost cheerleading with bunches of gum leaves in their hands. The welcome smoke that burnt in the small fire that evening wafted over me, and I gratefully accepted its blessing, and with it the commitment to respect Bunjil's lore.

Phillip enters the smoke and scoops it over himself. The droplets of water remaining on my body evaporate to dryness in the heat. The story of the river and the people who gravitate to its banks is still being written. What thread might I add to its myth?

In this small ceremony, I ask in my own way to be accepted into its narrative — to find my place on its misty banks.

The backbone of myth is knowledge. With knowledge comes understanding. And from the seeds of understanding, belonging can grow. Belonging, as the Shorter Oxford Dictionary says, is to 'not be out of place'. I'm ready to be *in* place, to be *within* place, to have place within me. It's time to apprentice myself again to a single circumference of life, to get to know its inhabitants, its moods and cycles and seasons.

I'm ready to belong again to somewhere rather than everywhere. And there's one very fast and efficient way to do that. I need to find my sit spot.

The next morning, I wake, sling my binoculars around my neck, and step outside. Out of the corner of my eye, I catch a particular flutter of wings in the camellia and immediately recognise it as a grey fantail. It issues forth a long, high-pitched trill, cinches in its waist, and proceeds to flaunt its tail feathers towards me in outrageous flirtation. I giggle and call a hello. I've already picked up on the presence of a few grey fantails in the vicinity, and potentially a rufous fantail too. My bird radar has been active as I've been settling into my room, eager to register my cohabitants. I've mentally logged about twenty species so far and have a few other hunches.

The few brick steps that lead down to the dirt path are cold underfoot. I reach the threshold of 'the forest' — the steep bank of casuarinas, wattles, blackberry, an ancient old walnut, and a tall silky oak bursting with spectacular orange flowers. The fantail hops from branch to branch behind me. The winding dirt trail levels out onto river flat. To the west, the neighbour's guest bungalow is obscured by a stand of elm and ash, while to the east a broken fence barely demarcates the other neighbour's yard, which is dotted with young eucalypt saplings and graceful tufts of poa grass. The riparian strip on our land features two old river red gums, their trunks needing two of me to hug their girth.

I pause and look around, a little nervous, as if I'm about to meet someone significant. I am, I remind myself. This place is going to be a significant relationship — the place that I commit to getting to know above all others, listening in on all its intimate conversations, staying loyal to even when the storms and rain roll in, keep coming back to even when I want for something fresh and excitingly new. This will be the place I apprentice to, knowing it will show me just as much about myself as about it, and hopefully, if I'm attentive enough, offer me a seat of belonging on this riverbank.

Walking between the two river red gums, I feel like I'm passing between two guardians. On impulse, I get down on hands and knees and crawl under the low-hanging branches of a small paperbark. Clearing some dead twigs, I wedge my back against the narrow trunk and tuck up cross-legged on top of the thick tradescantia. If I stretched out my feet, they would dangle over the bank with a few metres' drop into the water below. A large stag broken off at half height is set deep in the mud to one side, its root system creating a little village of reeds and rushes and small eddies below me. A breath of bubbles rises to the surface from within one of the root wells. Eel? Carp? Turtle? I'm curious. I like being down here, close to the mud. Across the river are the sun-bleached limbs of the fallen tree where Phillip and I perched. Nine Pacific black ducks swim upstream and fan out into roosting sites along the branches. I can see the sky, the river, the bush opposite, the thickets either side of me. Usually, I would hunt for a view, but there's something about being hidden that feels good. I can pretty much only be seen from the water. Maybe it's because life in the city falls so intensely under the human gaze; there's barely a pocket outside the four walls of one's house where human witness is not assumed. I hadn't realised the effort that being subject to such observation requires until I crawled into this nook and felt my body let go in some subtle way. There are ways I know myself only when surrounded solely by the mirrors of the other-than-human. My sit spot is teaching me already! If it is indeed my spot. Let's see.

I close my eyes and tune into sound. The loudest tones are the drone of the Eastern Freeway and the wheeling riot of rainbow lorikeets. I dial those two down in order to catch the quieter ones. Silvereyes, companion calling. I imagine them up the slope picking off insects within the corky rolls of silky-oak bark. A raven calls from the east.

A sudden impulse to open my eyes. Greeting me is a bright blue head, popping cheekily out from one of the hollows in the stag. A rainbow lorikeet. Another pops out of the same hollow, nestling in close to its mate. Against the bone white of the timber, the primary red, blue, and green feathers are a circus for the eyes. The birds bob their heads in my direction as if welcoming me to the neighbourhood. 'Well hello, you two!' I call out in greeting. 'Looking beautiful today.' I wonder if they're nesting. One of the things I love most about keeping still for long enough in nature is seeing critters in these unguarded moments. A feeling of empathy sweeps through me. This is the start of it, I think with a sigh. The inevitable falling in love. The feeling is laced with as much fear as joy.

I've had three significant sit spots in my life, and three significant romantic relationships. Leaving each of them has been accompanied by not small degrees of heartbreak. Like the human relationships, I can still taste the flavour of these places as if they were with me now. Each one with its own distinct personality, its own scent, its own cycle of allurement, plateauing and deepening. Like me, the places and people that I have loved grew into themselves more by being the object of attention, by the knowing of themselves through the lens that I offered. It was a reciprocal agreement.

My first sit spot was also in a city. I was living in another grand three-storey share house called 'Sunnyside' — a former rectory on top of the hill in Newcastle that a friend's mother owned; sunny not only because of the yellow paint that caught the morning sun as it rose over the ocean, but in disposition too. Out the back gate was the 'stairway to heaven' that led up to the city's Anglican cathedral along with a grassy park and small olive grove that overlooked the busy street below. The olive grove became my first sit spot, back in the days before I owned binoculars or even a smartphone for my bird app. I made up for the lack of experience, field guides, and natural diversity with enthusiasm. My flatmate Jo and I committed to doing a sit spot every morning and sharing stories over morning porridge. For the better part of a year, I sat most mornings with my back against an olive tree watching a tiny patch of wildish ground go about its business while the city buzzed and hummed all around. My 'teacher' at this sit spot came in the form of the pied butcherbird. It was actually the only bird I saw there apart from Indian mynas, but it was all I needed. Over four seasons, I watched a pair of butcherbirds build their nest twig by twig in the olive opposite me, lay their eggs, guard them from danger, and eventually welcome the first cracking of the shell. The peaceful sit soon became an ear-piercing cacophony of hungry begging calls, the two parents exhausted in their bid to feed the hungry beaks. The chicks rapidly grew big and strong, muscling each other out of the nest until they abandoned it altogether. Their awkward flying lessons between the branches mirrored the naturalist training wheels I was moving about on. Finally, they flew the olive grove, as did their parents shortly after. It wasn't until they had gone that I realised how much I would miss them, how much a part of the family I felt like I had become. Each sit spot since then has also been characterised by a particular bird that captures my curiosity and has me wanting to return. I wonder what bird might show up here.

Just as I have that thought, I hear a familiar shrill piping and look up to see an eastern spinebill in the neighbouring wattle. 'Heya, spiney!' I call out, accustomed to voicing out loud my greetings. This was my guide at my bush-year sit spot. I remember well my first encounter with spiney. After a couple of weeks of sit spot 'auditions', I had finally settled on a place on the upturned root ball of a giant tree. My back rested against the thick dried roots that rose above me like a woven wooden throne. The spot was on the edge of where drier forest merged with the wetter riparian zone, and on the juncture of two old fire trails. It positively teemed with life. Spiney flew into the shrubby banksia on day one of this sit spot and sang in the way I do when no one is watching. So close I didn't need the binoculars to see its long black hooked beak and deep, rusty orange pitted against black and white necklaces. It hovered like a hummingbird at the tip of the banksia flower, gently prying for its nectar. Not a rare and endangered species, I soon found out, but a common honeyeater, yet this common bird led me deep into the heart of the mysteries of my chosen area. My heart suddenly aches as I linger in the memories of that place, which grew to be like an extension of me. I'm haunted again by the knowing that I didn't ever say goodbye to it. And that it might not have let go of me either.

The truth is I haven't committed to a sit spot in the same way since I left the residency seven years ago. That year, my primary relationships were with the creatures at my sit spot — the rufous fantails, black cockatoos, small-eyed snakes, swamp wallabies, white-cheeked honeyeaters, wonga pigeons, and brush-tailed possums. Trees were friends; nests and southerly clouds and fox trails were mapped in my mind so vividly that they had become part of my own neural pathways. The map of myself had merged with the map of the land. Uprooting myself left a wariness inside. And now, this place is tugging at my heartstrings.

Kek Kek Kek Kek. There's that raptor again. I crane my neck in the direction of the eastern canopy and just catch sight of it flying in the same direction and pattern as the other day. Maybe this is my doorway bird! I feel like I'm at my first day at school, sitting up straight, sharpening my pencils, ruling my margins. All ears for the scratches and shufflings around me. Including the drone of the highway.

Despite the wildness of this patch I sit hidden within, the reality is I'm sandwiched between major arterial roads, all leading to and from the high-rise city centre not five kilometres away. Plastic refuse collects thick as carpet within the limbs of the half-submerged fallen tree, a reminder of takeaway suburbia that extends for kilometres in all four directions. I'm in the thick of it.

I want to know this place, to discover its crevices and burrows, its mysteries and treasures. I want to know these strange wild others, get acquainted with their habits, idiosyncrasies, and diurnal doings; be privy to their routines and ruts, preferences and pleasures, flappings and flowerings, struggles and celebrations. I want to know how this wildish place interweaves with the civilised landscape within which it's nested, how they inform and support each other, how they inform and support me. *See the tree waving in your direction!* it says. *See the birds fly in low and close! See the berries to eat, see the sweet river to bathe in!*

The place is courting me, but will I accept?

Chapter 2

Riverhouse Nature Journal — 5.37 am — Monday morning. Kookaburra and butcherbird are first birds. Silvereyes come in. Can hear so much more before the lorikeets start at about 6 am. Boobook owl! Coming in from eastern neighbours. V exciting. Where does it live? 2 flying foxes head home. Flock of ducks madly flying downstream. Where from and to? Can hear something rustling in the bushes behind me. Could it be the fox? Blackbirds suddenly bust out of the tree at the river — were they sleeping there? Something climbing the tree on the eastern fence — possum? Also heard 1st fantail cuckoos coming from the west. Lilly pilly near the corner of Separation and Gillies St flowering.

'*Cooooo-eeeeeee!*' Oh, that's for me, I remember after a few seconds. I was absorbed in the beep and tweet back and forth of the thornbills and pondering just how textbook an example of a companion call it is — one of the five kinds of songbird calls; translated to human terms, it might equate to sending a text message to your friend in the next room every five seconds to make sure they were still there and thinking of you. In attachment theory, that would probably warrant a diagnosis of extreme anxious attachment. I'd be jittery too if I thought my mate could disappear into the beak of a raptor at any moment.

I pause for a moment before coooo-eeeeeing back, enjoying the parallel experience of being beckoned by a companion call from my own kind. The thought suddenly occurs to me that I too could be under observation by some other external and invisible force, just like I'm observing the thornbills. I sit up a little straighter at the thought of being the watched rather than the watcher. I look forward to the time when the sense of being the outsider looking on dissolves and sitting here feels as natural as the thornbills calling above. The wind sweeps through the trees and showers me with a sheet of water, adding to the constant pitter-patter on my raincoat. My jeans are already soaked through after forty-five minutes of sitting still in the drizzle.

I crawl out from under my tree and walk stiffly back up to the fire circle. Something rustles from a little way upstream, and I turn to watch Mel emerge from behind a clump of lomandra. In her brown corduroy trousers, standing a good ten centimetres shorter than me, and wearing a green rain cap pulled over her eyes, she appears a bit like a hobbit emerging from the mist. I smile as she approaches.

'Ohhhhh, so much happening this morning!' Mel says, stretching out her arms to indicate the breadth of action at her sit spot. Her excitement is contagious, and the excitable naturalist kid in me is jumping up and down on the spot, busting to share my stories.

'The lorikeets are going nuts,' I exclaim loudly to be heard over their racket.

'I know, I know!' Mel says, smiling broadly. 'And did you see the eastern spinebill calling from the top branches of the hawthorn?'

'I didn't see it, but I certainly heard it piping away,' I say. A glow of joy passes through me at the opportunity to share what is usually a solitary experience.

It's the fourth Thursday morning in a row that Mel and I have met at 7.00 am for a sit spot in my backyard prior to the home-school nature-play program that she runs in a park nearby. It's been great timing for me. Three months ago, I bought a smooth

A4 unlined book and titled it the Riverhouse Nature Journal, where I record jottings from my sit spots, including sketches and maps. It's a way of keeping myself motivated — by sharing the stories with my journal, I track my own questions and curiosities. I've also started an independent naturalist home-study program called Kamana, a handhold through the practices of learning to see the land with the eyes of belonging. It's naturalist training 101, but I'm enjoying adopting beginner's mind again and having the accountability of geeky workbook tasks to complete. Still, there are limits to the efficacy of solo study. After two months, my journal entries began shrinking in length and frequency. I've stopped tracking the numbers and directions of bird movement, started skipping consecutive days at my sit spot. A dangerous thought has crept in — maybe I've seen all there is to see. *After all*, this voice says, *it's just the city*. I've been giving myself the same sit-spot pep talk I give my students — 'love the one you're with' — but the thought is lingering. These weekly shared sit spots are part of my strategy to amp up the curiosity quotient.

A spontaneous call from my old friend Matt this week has also helped light the fire again. He's a bit of a sit-spot guru, having sat almost daily over the last five years in the same spot near his home on the New South Wales mid-north coast hinterland and pieced together a seasonal calendar that would surpass that of an ecologist. I had texted him a currawong question, and when he rang back he got the vibe that my sit-spot honeymoon period was waning. I told him a bit about my area, and he listened carefully before gently starting to ask some questions. 'Have all the flying foxes moved since summer? Are there still ripe blackberries? Could the fox be eating them?' I ummed and ahhhed and flipped through my journal.

'So you're telling me you've got a flying-fox colony downstream, a resident fox, possums galore, an unidentified raptor, a river, and you're not curious?' Matt asked bluntly. Point taken. Matt suggested that I come away with at least one question every sit spot and write it in my journal. It's been a great invitation, and I've been journaling at least five.

The unidentified raptor is the top of my mystery list, as is the whereabouts of fox. There's a distinct fox-sized trail that leads across the neighbour's broken fence parallel to the river. I've tried following it, but it peters out. I also followed the creature's musky scent trail one day to the base of one of our river red gums. *How large is its range?* I asked my journal. *Where is its den?*

Then there's the owl that sounded like a screeching woman one night, and the next morning Min showed me a dead brush-tail with small claw marks on its body. Could they be linked?

It's also the season for babies. I heard baby bird cheeps some-where in my room and then later found bird poo on my pillow. There's a baby myna bird outside my window and baby ringtail possums in a drey that I can see from the window of the spare room. At dusk, I watch the pile of soft fur wriggle and awaken. There was a baby blue-tongue lizard on the front pathway, its tongue a similar colour to the cobalt blue of a worm I found in the garden. Then there's the family of baby tiger snakes that Elizabeth found when she upturned the blow-up canoe. Their presence as a regularity was confirmed by the neighbours who told me their dog had died from a tiger-snake bite a few years ago. They expected me to look dismayed, but instead I barely hid my excitement.

I love that there's a deadly snake near my sit spot. It's the perfect goad to get out of my mind and into my senses. After all, the fear of being eaten on the savannahs by lions is what kept our ancestors in a state of continuous landscape tracking. They couldn't afford to be ruminating all the time or they might be picked off. Besides, I have a strong regard for tiger snakes. I almost stepped on one once, camouflaged as it was in the leaf litter of a bushwalking trail. I was days into a solo mountain hike with no phone or emergency beacon. My gaze fell on it a second before my foot would have made contact. I've wondered whether that might be the closest to death I've come. The incident propelled me into a heightened state of aliveness — part adrenaline and part reminder of the fragility of every moment. It's good to be prey sometimes rather than always predator. It's humbling.

Mel and I are soon sitting on the couch with steaming cups of chai and toast, our questions and stories tumbling over each other.

'What was the duck with the mottled brown chest?' I ask. 'A wood duck?'

'Yeah, I saw that, not sure, maybe a juvenile, and there was another maybe male with a darker head,' Mel says.

'And the moorhen picking through the rubbish on the far side. The ironbarks are a feast for the lorikeets. I wonder what will happen when they dry up?'

'I've been wondering that too. Do they migrate locally? Or is there always something flowering to keep them around here? And are they all rainbows lorikeets or are there other kinds here too?'

I take a long sip of chai and ponder the questions. They are almost as delicious as the buttery toast.

'Rainbows definitely, but I assume there's at least musk lorikeets around here too,' I say.

'Careful of those assumptions,' Mel says, laughing, and buries her face in the mug. I feel a flood of gratitude. Not many friends of mine want to get up early and watch birds in the rain with me.

'Hey, can I give you the tourist test?' I ask.

'The what?'

I run downstairs to retrieve my notes on the twenty-question quiz in the first few pages of the Kamana workbook. Named the 'tourist test', its questions are designed to give you an indication of how much you actually know about where you're living, like a belonging ratio. As I take Mel through the questions, I read back through my own answers.

What is the edible or medicinal plant nearest to you, and how can it help you this season? *Dandelion, liver cleansing.* What phase is the moon in, and where is it in the sky? *Quarter and ... ??* What direction is the wind blowing from? *South ... I think?* What is the nearest wild edible to your back door? *Wood sorrel.* What are five mammals that live within two hundred metres of your house? *Cat, dog, brush-tailed possum, ring-tailed possum, fox.* Draw a map of your backyard and indicate true north. *Well, north yes — but true north?* What are the birds telling you about the location of large carnivores in the area? *I witnessed the blackbirds outing the neighbour's cat with an alarm call.*

I wasn't doing too badly until I got to the water questions. Where is the closest natural drinkable water source to you, and how much water does it yield every day? Where does the water come from when I turn on the kitchen tap? Where does it go when I flush the toilet? *Blankety Blank.* I assumed my drinking water came from the mountains at the source of the Yarra, but I wasn't 100 per cent sure. The bigger wake-up call for me was the realisation that *I hadn't even asked the question.* I just took it for granted that water flowed from my tap like magic and then left again. It's not like the immediacy of dipping a bucket in a stream or tapping the side of a rainwater tank to ascertain its depth level. One needs to be intentional in the city about discerning the connections that support life.

Having grown up in Melbourne, Mel is much more informed on water but stumbles on the bird questions.

'It's a bit of a reality check, hey?' I say. 'So you can be my witness — I commit to finding out and seeing with my own eyes the places of my water's origin and destination.'

'Done. Hey maybe we can introduce this tomorrow?'

'Yes, let's!'

'Cooooo-eeeeeee!'

Elena emerges first from her sit spot down by the creek, dark Italian curls framing equally dark eyes and a pale pretty face. One hand holds heavy hiking boots swinging by the laces, her bare feet shining out against the dark clay ground. Her walk back is slow and somewhat mechanical, carrying an air of both beauty and sadness.

'Bare feet … nice,' I say as Elena approaches.

'Yes,' she says stiffly with a half-smile that's quickly checked, as though negotiating with a gatekeeper on her shoulder.

'See much?' I ask.

'There were some birds around, not sure what they were. But my mind was still so busy.'

I remember back to our short phone conversation last week. Signing up for this program meant negotiating down to four days a week Elena's role as project coordinator in a university library, something that by the sounds of it took a lot of courage even to ask for. It's part of what I ascertain is a last-ditch attempt to try and dig her way out of a deep depression that's had her in its grip for the last few years.

Tas pops up next, so tall that, once standing, he's a head height above the shrubbery. He looks reluctant to come back in, lingering at the creek edge with a quality of reverent stillness. Eventually, he turns and begins to pick his way through the tree violets and acacias.

I first met New Zealand–born Tas last year, when he brought his two primary-school-aged boys along to our first Wild by Nature Village Camp — a weeklong nature-connection camp-out for families. He had grown concerned about his boys' increasing wariness of nature and wanted to get them out in it. Tas watched from the sidelines the first few days, his boys equally cautious.

One morning, I took the two kids on a wander down by the creek, where we found ochre rocks and dot-painted our faces. A golden whistler flew in low, and my gaze followed it to a nest in a low fork of a melaleuca. Together we crouched in hiding and waited, the boys barely containing their excitement as they witnessed the striking bird flutter about the nest not metres from us. We moved in a little closer and my co-mentor Josh hoisted the boys up so they could peer over the rim of the nest to see two spotted turquoise eggs not much bigger than a marble.

'Whoa!' they whispered, eyes like dinner plates.

I was equally transfixed, so strange and beautiful were the eggs resting in their basket of sticks and moss. It was as if an artist had suddenly pulled us into her surreal life painting. My awe was amplified by seeing it afresh through these children's eyes. It was a rite-of-passage moment, I thought standing there while the two youngsters glanced back and forth between me and the nest with a look of unabashed awe: the moment when one first witnesses eggs lying in a nest. It's an image so simple and yet powerful in its symbolism of the fragility and preciousness of life, the very miracle of creation. The bird flew in again, perturbed by our presence. We could clearly see the distinct rings of yellow, black, and white on its bib as it perched protectively on the edge of the nest.

'What bird is it?' I said, feigning ignorance so they could be inspired to solve the mystery, which they did, with a flurry of excitement when they turned to the page in the field guide I carried in my backpack. That bird would never again be a stranger to them, but more importantly nor would the memory of the vitality they felt at being woven into one of nature's mysteries. I watched them run towards Tas on return from the walk, tumbling over each other to tell the story. Something woke up in Tas that week too. It hadn't just been his kids that needed rewilding. He organised Fridays off work from his osteopathic practice so he can come along now, and his leather moccasins veritably glide across the grass towards us.

Michelle takes time to appear, despite the fact her sit spot was so close I could have almost whispered her a personal cooee. In one hand, she carries her journal, while the other flicks her shoulder-length brown hair back. I'm hoping the delayed return indicates a positive experience. After everyone else had drifted off, Michelle hung back to talk to me, divulging that she was fearful of sitting alone in the park. At first, I was confused and had to enquire further to understand that it was stranger danger that concerned her. 'Even in the daytime?' I asked, trying to mask my surprise, to which she nodded. Michelle has travelled across town to join the program from one of the western bayside suburbs, where she and her husband have just bought a house. She told me her back gate opened onto an open field and wetland, which she is struggling to both connect to and feel safe in alone — which was part of her motivation for coming along to these Fridays.

Before chatting to Michelle, I hadn't considered that fear of urban parklands could be one of the barriers to connecting with nature. It was good information. I reflected on my long walks down the river, where I could well pass only a few other walkers. Only after dark would I start to feel edgy. Then again, I had encountered my fair share of terror when I first started camping out alone. It took years to feel completely comfortable wandering around the forest at night. And I guess sitting alone in nature even in a park is not familiar to most people. Fear is a natural response to unfamiliarity — the reptilian brain doing its job of going a bit crazy in a new situation. What was that rustle? It has to be a snake. And that dark shadow over there? A shady character spying on me. What was that on my back? It must be a spider. The greatest fear, though, is fear itself, I learnt when I started to confront my fear of the dark. Telling myself there was nothing to be scared of didn't help much at all. What did was repeatedly putting myself in the fearful situation and having the contradictory experience that I wasn't mauled by dogs or lost in the woods. Everyone's edge is different.

The daughter of Croatian migrant parents, Michelle had become the subject of a few family jokes about signing up to 'this rewild thing', but the way she relayed it told me more about her determination than anything else. I had encouraged Michelle to find a spot within sight of camp central — 'Start close in,' I told her.

'How did you go?' I ask as she approaches, and in reply Michelle gives me two thumbs up and a broad smile. A small step in the right direction. A few other participants follow in Michelle's wake.

A scuffle of movement up the slope reveals Lucy, a semi-retired paediatrician. Her lithe figure moves deftly over the fallen logs on the hill and down to the grassy clearing. She bounces towards us with a mischievous grin, looking back behind her as if privy to an invisible group member. Despite her short grey hair, there's something decidedly childlike about her. 'I could have stayed there all day!' she says in a clipped English accent, her blue eyes sparkling. 'What's happening next?'

Mel and I look at each other and laugh.

The last to return is Laura, who creeps up silently behind us from the wilder side of the wetland. Her face is moistened with what I guess to be a splash in the shallows, fine water droplets clinging to her ringlets.

'So many birds, so much life,' she says with a thick Spanish accent. She looks at me with a kind of quiet, interrogating gaze but tempered with layers of warmth. I already feel a kinship with this woman; she has an air of wolfishness, a self-containment, the lines of communication between her and the natural world humming with life. As I watch her squat on the outside of the group, I wonder what's drawing her here, where her learning edge will be.

We count heads. A full dozen accounted for. Mel helps organise everyone into pairs and invites them to share sit-spot stories with each other. Soon they're chatting away like gaggling geese. It's a back and forth, this sit spot and the storytelling, like another companion call — completing halves of the same practice. Sit. Listen. Observe. Question. Share. Question. Get super curious. Repeat. One follows the other in a mutually supportive cycle. We briefed each pair to not only listen to each other, but to ask questions that take their partner to the edges of their awareness and show them as much about what they missed as what they did notice. Only, what I'm coming to realise for myself is that the story swap is not just about invigorating curiosity but about bedding down all the sensory information received so that it becomes part of a new neural map.

Mel and I partner up to swap our stories and end up talking about the location we've chosen to host our Rewild Friday pilot program in. Today is the first day of a weekly class in nature connection and wild-living skills within the city's bushlands and parks. We're at a thirty-three-hectare bushland reserve in the middle of the north-east suburbs that was once a forty-metre-deep quarry and municipal tip. The bubbling stone creek that runs through it empties into the Yarra not far upstream from my share house. We chose it as our primary location due to its feeling of wildishness, and because the ranger, Dave, has agreed that we can have small fires here between May and September, which is a huge boon. We plan on basing ourselves here so we can get to know one area well, and only take off on excursion once a month.

The idea for Rewild Friday sprung from a conversation I had with Mel where I was dreaming up a regular learning space for naturalist-tracker types. Nice idea, Mel said, but we both ultimately agreed that, given the limitations on people's time, it would likely turn into us teaching for free. I grappled with the concept of monetising something so simple and second nature to me. No one should have to buy it. Is it ethical to commercialise something so fundamental? Something that nature gives freely? But if I didn't ask for a monetary exchange, I would probably burn out. I need to pay bills too.

I'm running a business, employing others, paying taxes. Teaching and guiding is one stream of income, alongside writing, speaking gigs, and individual mentoring. A few times a year, I take people out bush on longer nature-based personal-growth immersions such as Vision Quest. The boundaries between work and personal life are thin. Even coupling the word 'business' with what I offer feels strange. Still, we decided to put the Rewild offering out there. I secretly wondered whether anyone would book in. To my surprise, the first term almost booked out. And now here we are.

'Any stories?' I invite, bringing everyone back to the larger group.

'I found some diggings,' offers Laura, showing the shape of them with her hands. 'I wonder who made them.'

'I tried tuning into the quietest sound as you suggested,' says Tas, 'and realised there's all these bird calls that I never pick up on. But yeah, it was amazing to notice how much I don't usually notice. It's a bit overwhelming actually.'

'To be honest, I found it hard to keep sitting,' says Lucy with a smile. 'I just had the urge to keep exploring uphill but wasn't sure that was allowed.' The rest of the group laughs.

'Allowed?' says Mel. 'There's no rules here. If you have an impulse to explore, then go for it!'

'Just be back before dark,' I add, laughing.

Mel gave a similar answer to Michelle at the start of the day when she asked to use the bathroom. Michelle looked a bit shocked at the answer of 'do whatever you need to do', like she was noticing for the first time that the cage door was actually open. We don't want obedient students; we want sovereign creatures following their own scent trail.

The remainder of the morning, we focus on basic natural movement and sensory awareness — walking quiet and silent like a fox with bare feet sensing the ground, seeing in expanded 'owl' eyes. It's the beginning of retraining the senses to be powerful receptors again, soaking like a sponge as much information as possible about the landscape rather than the very limited range of perception we are accustomed to. The sun finally comes out, and we spread picnic rugs for lunch. Some of the group take the opportunity to stretch out in the sun like lizards on the edge of the wetland.

Lucy approaches. 'Can I share something with you?' she asks shyly. I indicate in the affirmative and make space on the rug. She perches on the edge, looking pensive, and I notice her striking cheekbones.

'I just feel I want to share why I'm here,' she says, playing with something that looks like a dillybag in her lap. I nod encouragement.

'A few months ago, I had an impulse to find my bilum — the string bag that I made with the women of the New Guinea village I was living in during my early twenties. I hadn't thought about in years, and one morning I woke up and knew I needed to find it. I thought it might have got lost in the divorce and move, but I found it in the bottom of a cupboard, in mothballs.' Lucy holds up the piece in her lap with the tenderness one might a baby. I see now it's an expertly made, finely woven string bag.

'Wow, you made this?' I say, reaching out to touch the bag gently. It's soft and supple.

'It's made from cuscus fur that I bought at the market. The fur together with the blue thread makes it a ceremonial bilum. I sometimes wonder how I had time to make it with a full-time job and an early-morning run through the mountains. I burst into tears as soon as it was in my hands again, and I didn't stop crying for three days. I don't know how to put it into words. It was like all of the past and the future were in that very moment. There was an overwhelming sense of something returned to me, something important that had been locked away all these years as I worked hard and raised a family, and I hadn't even realised I was missing.'

'Could you say more about the feeling of what that thing is?' I ask gently.

'I don't really know, but I'm calling it my wild self, without really knowing what that means. It's strange, the words "learn tracking" came to me, and I had to google it to even find out what that meant. That's how I found you and this program.'

Goosebumps appear on my arms. The mysterious stories of how nature and psyche are bundled up together never fail to amaze me. It's a bundle that longs to be unwrapped, even for Lucy at the age of sixty-six. I look at the bilum in her hands and imagine the radically different life of the women who taught its crafting, in a tight-knit subsistence community in the misty tropical mountains, compared to that of Lucy, alone in an apartment in the city. The string they wove for her is tugging her back. Might they be remembering her too?

'And so you're here in part to reclaim what has been in mothballs all these years,' I reflect back. 'Or at least find out a bit more about what that is.'

'Yes, and I have no idea what that means yet,' Lucy laughs with tears in her eyes. 'First step is just turning up.'

Suddenly, her demeanour changes and she jumps up. 'That's all, just wanted you to know.' She smiles before turning away.

My feeling is that contacting the bilum has awakened much more than Lucy realises. What's clear is that she's bursting out of some kind of holding pattern with the rip-roaring holler of someone who has suddenly realised that she has an entire life within her as yet unlived and is clawing at the door to be let out.

What you seek is seeking you, I think to myself, recalling Rumi's words as I watch her walk away to join the others in the sun.

The end of the day. Mel and I plonk ourselves down on a huge fallen log to rest.

'Phew, well, that seemed to go well,' Mel says, her laugh belying the understatement of her words. The group we just farewelled were almost unrecognisable from the one that gathered this morning — bright-eyed, with mud between their toes, smiling broadly as they called out a 'see you next week'. Elena has lingered too, and we watch her wander barefoot to stand at the edge of the phragmites by the pond.

For our final exercise, Mel and I had created a circle of bark that everyone stepped into one by one to speak their intentions for the program. Elena had surprised me with her statement, announcing to everyone that she had 'injured instincts' and was here to heal them. It struck me as such an accurate and poetic way of describing the malady that is written in her every gesture. Lucy went last, entering the circle and issuing a loud wolf howl, much to the surprised amusement and resonance of the others, who realised this more than anything linguistic spoke for them too. The sound issuing from the lungs of this woman who had been silent for so long gave stirring voice to the wild self that she is now tracking and apparently is tracking her.

Injurious indeed has the rationally human-centred culture been to our instincts, to the intuitive, imaginative, creative ways of knowing, to the part of us that longs to wade through waist-high mud and dive headlong into freezing mountain lakes, that tunes its animal senses to the smell of the southerly before a cloud has graced the sky, notices the belly graze of a lizard in the sand at the back door, and mingles minds with the screech of the cockatoos at dusk. That self is the sensuous, erotic, instinctive, howling, laughing, dancing, loving, crying, enchanted, spontaneous, loud, belly on the earth, sighing, yawning, screeching, singing one.

It's as this self that we feel the bone-deep relief of being completely at home in our expressive, sensuous animal bodies, in our cataclysmic emotions, and in our intense love for life in all its physical forms. Here, thinking gives way to feeling and sensing. Concepts shrivel up under the weight of full-bodied sensations, the desire to press our bones against granite and our flesh into clover, to reach up to pick sweet berries from the bush and blacken our hands with charcoal from the shards of last night's fire. This is the part of us full of devotional love for the world. Far from apologetic, this part of our wholeness feels fully entitled to take up space — to unconditionally belong. It's as David Abram says, 'Becoming earth. Becoming animal. Becoming, in this manner, fully human.'

Sharon Blackie speaks of this archetype as the Wild One, the part of us that naturally moves towards earth-based wisdom and a sense of kinship with the rest of the natural world. More broadly, it is, she says, 'an expression of the nature of our own unconscious, a portal into the depths and richness of our inner landscape'.

It is this Wild One that I went in search of during my bush retreat, intent on reclaiming my own fierce and loving attunement to the wilds both outside of me and within. Eschewing words for a time, I looked instead to a language older than words: the purring of creek, scratching of claw, and haunting melody of wind through storm-blown treetops. Instead of sitting in any form of submission to a culture intent on destroying the wild, I supped on it daily, embodying it in every subtle posture, feasting on its wild berries and dark, wooded dreams. And the wild responded with equal ferocity, stripping me back to raw essence, ravishing me with its elements until I couldn't ever lose the capacity to see the world through at least one yellowing raptorial eye.

But here, with Lucy in her apartment with her bilum, Elena in her office job, Tas with his busy family life, and all the others tied to all the numerous buzzing strings of the city, what can this one day a week and these practices really offer? How deeply can they go? What happens when you wake the sleeping giant of biological ancestry from its slumber? Can the wild and the urban find a happy marriage in a normal city life?

I've suggested they find their own sit spots close to home. My hope is that the nature-connection learning cycle will grip them with fascination and curiosity like it did me, but I really don't know. Maybe they won't be able find a sit spot they like, or maybe they'll never find time to go there. Maybe the six hours on a Friday will just be a kind of nature-inspired yogic out-breath to defrag their otherwise hectic life. I'm really not interested in providing a palliative experience. I'm interested in shifting the very foundations. Rewilding isn't just about nature connection, it's about reclaiming our own metaphorical bilums — that wild core of ourselves whose passionate teeth and claws would soon unpick the current cultural fabric in favour of something fire-breathing and earth-real.

I watch Elena as she walks back to her car, and my heart suddenly goes out to her. She's here following the single thread that she can find some trust in — an impulse to be barefoot on the earth, to be caressed by both sunlight and moonshine, to carry dirt in her hands and the sound of birdsong in her ears. To apprentice herself to those who are instinctive and life-filled. And in this movement, a quiet, desperate prayer that she will find this within herself. I wish that for her. Time will tell.

Chapter 3

I press the buzzer on the wall of the red-brick block of flats for the third time and wait some seconds. Still no answer. Hmmm. Weird. This is the time we had arranged. Maybe I've got the address wrong. It's certainly not what I was expecting for this famed wild one of the suburbs. Perhaps he's gone walkabout. Just as I'm turning around, I hear an opening of doors, and a young man appears. His dishevelled dark curly hair, prominent forehead, and dark eyes speak of a Mediterranean heritage, and I place him at about twenty years of age. He leaps down the last set of stairs in frayed Dunlop Volleys, a pair of black cotton shorts, and a dark-blue woollen jumper with snow-like patterns around the neckline that reminds me of one I owned as a child. He carries nothing but a pair of binoculars around his neck. No water, I note.

'Gio,' he says, offering me his hand, his voice much deeper than I anticipated.

My friend Fam told me first about Gio, describing him as a self-taught naturalist kid who lived with his mum in a flat and spent his time up trees or down burrows in his neighbourhood. He worked casually at the local environmental-education centre, but his propensity for no-shows made gainful employment difficult. The few others I knew who had met Gio spoke of him almost mythically, as an elusive creature who preferred the company of the furred and feathered ones on the suburban margins to that of humans. It reminded me of the European tale of the 'boy-wolf' who grew up as one of the pack until he was 'rescued' back to the village, where he lived uneasily, padding around the fence lines and looking towards the distant hills, never losing the scent of his former life in the mountains.

It took quite a bit of dogged tracking to pin Gio down via technology. I was surprised when he finally returned my text message and gave me his address in one of the most population-dense areas of Melbourne. That's when my curiosity was really piqued. His street is treeless, landlocked, and wall-to-wall with 1970s apartments and the odd millionaire's mansion. Surely an unlikely place for a boy-wolf to grow up in.

'Ready to go?' he asks rhetorically, already pacing down the street ahead of me.

'Ah, yeah, ready,' I say, slinging my small pack over my shoulder.

Gio's claim to fame is a twenty-four-hour 'Bio Blitz', where he set out to prove that there were more native species living in our suburbs than there are species in the Melbourne zoo. The list far surpassed that threshold, and Gio documented more species in his local area than ecologists ever had, including a rare eel in the stormwater drain.

I was hoping to get a little bit of context about how he came to be sloshing about in drains looking for eels despite what seems like no cultural or familial modelling, but it's clear that past-tense conversation isn't going to be compatible with the whistlestop tour he has planned for us. I catch up to him just before we cross a major four-lane intersection, jump the fence, and walk straight out onto an immaculate grass fairway. I look around to see if any golfers are teeing off in our direction. Gio, however, seems not the least bit concerned, and strides through the course without clubs as if it was his backyard. There's no arrogance, though, more an endearing naivety and untrammelled enthusiasm.

'Musk lorikeet, masked lapwing, striated thornbill ...' He's rattling off names faster than I can follow his pointing finger. I curse the fact I didn't bring a notebook, and instead open up the back cover of Tim Low's wild-food book I brought with me and start scribbling.

We come to an artificial lake in the middle of the ninth hole, and I squat next to Gio at the water's edge as he talks me through the breeding cycle of the grebes, geese, and ducks.

'See, there's the Australasian grebe — they eat their own feathers and feed them to their young to prevent injury when swallowing fish bones.' I bring my binoculars up to my eyes to check out the rubber-ducky-sized waterbird. It's clear he's giving me a potted version of the much greater breadth of knowledge he holds.

'How did you get so knowledgeable?' I ask, fascinated. 'What motivated you?'

'I just spent a lot of time outside checking things out,' says Gio. My why questions seem to be landing as absurdities to him. As if he can't fathom not doing it. Turning over the rock of nature's workings is just what he does.

'Come on, I want to show you my latest project,' he says, pointing over at a stand of eucalypts that, as we approach, I notice almost all have nest boxes. Gio walks confidently over to one tree and begins scaling it as if he had glue on his feet. When he reaches the height of a nest box, he peers in.

'This one's empty. But not long ago, it had a family of eastern rosellas,' he says, jumping to the ground and bounding over to the next tree. My jaw slowly drops as I piece together the story.

'Eastern rosellas had pretty much disappeared from around here. So I thought I'd do something about it,' he says nonchalantly, his feet gripping the trunk of the tree as he hauls himself up to the first branch. 'See, only they could nest inside this box.'

I crane my head up to look at him. I'm amazed. It must have taken seriously detailed observation to work out how to design a nest box for one specific parrot. Councils would scratch their heads for years before contemplating such a rehabilitation project. And it's been entirely self-initiated. I also learn that some of the nest boxes he's mounted are for microbats, owls, and kingfishers.

'Does anyone have any idea what you're doing here?' I ask, but he brushes the question off in a mixture of humility and disinterest. Looking up at this boy, who is barely twenty, with his head in a nest box, I wonder whether I'm in the company of someone who will one day be known as a national environmental hero. This grassroots apprenticeship is the kind of dedication that forms the likes of a Jane Goodall.

The feeling of trailing behind an unsung hero colours the remainder of the walk. We head to Gio's local waterway — Elster Creek — which holds the dubious fame of being the most polluted waterway in the state. Its concreted banks are a far cry from the river I live on, and for most would appear to be a stormwater drain, but Gio lights up as we get closer, jumping down into the ankle-deep water and squatting to examine something moving just shy of the concrete bottom. I wonder what he's seen.

'I've found short-finned eels in here before,' he says.

'What else have you found?'

'Ahhh, blue-spotted gobies, snake-necked turtles,' Gio says, trailing off when something else catches his interest. I'd heard from Fam that Gio had recently photographed a marine spider in the intertidal zone — the second recorded sighting in Victoria, the first being in 1902. I'm not sure if it's ironic or poetic that perhaps the country's sharpest young naturalist has cut his teeth on the biodiversity of the state's most altered waterway. It's either a symbol of hopefulness, resilience, or both.

Gio traverses the canal with the familiarity of a river rat, tossing out plastic refuse as he goes, his Volleys gripping the sides as if purpose made for the landscape. At a pedestrian bridge, Gio points out a blackboard mounted on one end, titled 'nature observations', which he set up for locals to record their findings. A public nature journal — what an awesome idea! 'White-faced heron' is scrawled in chalk with a date.

In some sections, the creek opens out to areas of mature trees, Gio pointing to the ones with hollows that boobook owls are currently occupying. I thought Fam was exaggerating when she told me that Gio knew the area's individual owls by sight. I believe her now. He has an otherworldly quality that feels akin to these out-of-the-way and night-world places he frequents. In naturalist language, they are known as human eddies — overgrown thickets, drains, tree heights — any pocket of nature that is for the most part invisible to humans intent on human business. But for rarities like Gio, they're a trove of treasures and mysteries.

Gio jumps up to take hold of a low hanging branch and swings back to the ground, glancing up as an ibis flies over. I'm finding him more fascinating the longer I spend with him. But what is it exactly? He looks back at me with the briefest of polite non-verbal check-ins. And then it comes to me. It's because he's one of the few white folk I've met who feels black. And the first one in a city. Yep, it's that. He feels indigenous, as if he's encountering the world through the eyes of one who belongs. The word fits even though I know he doesn't hold Aboriginal ancestry. *What do I mean by it here?* I ponder.

Our colonial histories have led to situations where descendants of the prior inhabitants, the peoples displaced by colonial incursions and settlements, have the first, and many argue the only, claim to indigenous status. I pricked up my ears once when I heard someone speak of three different kinds of indigeneity: terrestrial, ancestral, and place. Terrestrial indigeneity reminds us that at the very essence we are all born of the same earth and linked through our lineages back to the earliest humans. In this way, no matter what our heritage, where we live, or how connected we might be, we are all indigenous to the earth. Ancestral indigeneity refers to a historical continuity with pre-invasion and pre-colonial societies — the primary way we understand custodianship and speak of the sovereignty of this land having never been ceded by the First Nations. The kind of indigeneity that relates to place is less easily defined, more subjective. To be indigenous to place points to the experience of a location-specific belonging — the cultivation of deep relatedness and connection to one particular landscape in a way that is not dependent on birth or lineage.

Mythologist and bard Martin Shaw, who lived in a yurt through four English winters with a similar question on his tongue, asks us to also consider the difference between being *from* a place and being *of* a place. People can be from a place, he says, and be utterly oblivious to it. Being of a place means 'not talking *about* a place but *with* a place'.

'To be *of* means to be *in*. To have traded endless possibility for something specific. That over the slow recess of time you become that part of the land that temporarily abides in human form,' he says. It's to 'hunker down as a servant to the ruminations of the specific valley, little gritty vegetable patch, or swampy acre of abandoned field that has laid its breath on the back of your neck'.

Instead of the race and place debate, Shaw suggests we 'keep open to the fact that whatever your age is, there could be a place that wants to reach out and say, "Hey, stop. I've got something I want to say to you." And you cannot hear it until you slow down, till you stop multitasking, till you stop looking only at what is a foot and a half in front of you, and pay attention to all those trembling mysteries of that place that we just spoke of.'

Growing up between Melbourne and his Indigenous home country in Western Cape York, Tyson Yunkaporta writes in *Sand Talk* of the universal struggle with displacement: '[We are all] refugees severed not only from land, but from the sheer genius that comes from belonging in symbiotic relationship to it.'

Tyson's radical invitation into the world of the Dreaming asks us to look at how contemporary life diverged from patterns of creation and how Indigenous thinking can 'save the world'. In one interview, he is unequivocal on the imperative to reclaim from our own earth-bonded fragments a belonging to place.

'Indigenous is baseline human who you are. It's your factory settings. You can be taken away from that and domesticated into some kind of shrivelled, ridiculous other kind of person. But that was who you were born as.

'I keep telling people … don't be mining the margins for Indigenous wisdom. You don't need to be looking for some exotic other. If you really do look into yourself, you'll find fragments of that. And each fragment contains the pattern of the whole, and the whole can be extrapolated out from any fragment. So, you find your own way and your own stories.

'In Aboriginal Australia, our Elders tell us stories, ancient narratives to show us that *if you don't move with the land, the land will move you.*'

Gio has claimed indigeneity to place — to the concrete-lined Elster Creek and surrounds. He's indigenous not in the way he might be if his ancestors' bones were buried here, but in the way he has allowed himself to be claimed by, and in turn claim, this place. Gio feels like he's *with* this place, hunkered down as a servant to this gritty city pocket. He has apprenticed himself wholly to this place, swapped possibilities of greener pastures for specificity, learnt to speak the land's language through its tracks and signs, interpreting the stories they tell. Rather than being reductionist or dominating in his study, Gio flows with the movement of the place, with a similar 'dignified agency' that Yunkaporta speaks of when one finds access to the realisation of one's 'true status as a single node in a cooperative network'. He is a participant rather than an observer, part of the conversation rather than eavesdropper, as much a part of the place as the animals he tracks. The place now radiates through him, becoming wilder as a natural extension of his own rewilding — the birds returning, the eels celebrated, the spiders noticed. He is caretaker, storyteller as well as storymaker, protector. Following in his footsteps, I feel like I'm following a contemporary songline, one that he and the land are co-creating.

As we meander along the last stretch of the creek, a feeling of nostalgia creeps in. Gio's figure morphs, and for a few seconds it's the fire trail to my sit spot that I'm walking on; in front, not Gio but a younger me in khaki cut-offs and a black singlet top, watching for the collared sparrowhawk's afternoon flight, sucking on the red cupped nectar of mountain devils. The image superimposes onto Elster Creek with Gio, and in that moment I realise I'm still holding onto a false dichotomy between real nature and city nature, still maintaining some belief in a hierarchy of value that asserts that I can't possibly find the depth of conversation with nature that I can in a more pristine landscape. But here is this kid, showing me that it's entirely possible, and the only thing stopping me are my own self-imposed limitations.

I hosted some colleagues at my place one afternoon recently and when inviting them to wander out on 'the land' noticed the words catch in my throat, as if it was disingenuous or hubris to extend that phrase to the suburban environment. Where do I think 'the land' begins and ends? Is 'the land' not also the paved streets of the CBD? Is it not the air and sky and the water droplets in the clouds and the planted street trees and feral pigeons? The caveats I held on what counts as 'the land', I realised, were subtle and many.

Gio and I cross a busy street and loop back to the last stretch of the creek as it empties out into the expanse of Port Phillip Bay. If Gio was alone, he might continue his wander underwater amongst the seahorses as he is apt to do, but instead he stands next to me in silence, kicks off his shoes, buries his toes in the sand, holds his shoulders proud and square to the ocean, and releases a deep sigh of satisfaction. He is home. I follow his gaze out to the horizon, inspired to see the land a little more through Gio's eyes.

'Yep, just like that, as if you're cutting into cheese,' Phillip says, instructing me on how to best split the big rounds of wood sawn off from the fallen tree down near the fire circle. I bring the splitter down with as much force as I can muster, and it cleaves off a perfect sized chip. 'Sweet!' I exclaim.

'You've been trying to tackle it from the inside, but it won't work, you need to start from the outer edge,' Phillip confirms. On the next go, the splitter gets wedged in the heartwood, and Phillip shows me how to turn the whole thing upside down and bring it down on the blunt side of the splitter to release it. I can't believe this has escaped my bushcraft education thus far. It's revolutionary. And gives me another excuse to chop wood, one of my favourite pastimes. A whiff of smoke passes over my head. Sparking up a fire is one of the first things we do on our Mondays, and we just feed it with offcuts as the day goes on. We both love having a smoky little flame ticking away as we potter. It's so comforting. Birds seem to come in closer, as if it smudges away some of our human separation.

I have been getting comments on how smoky I smell, though. I can't wash it off in the river anymore: the neighbours have taken away our use of the jetty due to skinny-dipper sightings. Oops! Given the steepness of the banks, I am essentially cut off from anything but observation of the river. I knocked on the door of our other neighbours, who have a postage-stamp-sized jetty, but they mumbled something about public liability. I couldn't believe it! I feel landlocked, like a duck that's told it can't get its feet wet. Min took over a chocolate cake as a peace offering, but so far they aren't budging on the decision.

This morning, Phillip and I planted native seedlings along the riverbank — correa, blackwood, hardenbergia, and, my special request, an apple-dumpling (billardiera) bush. Hopefully, it will yield some delicious fruit. I've also divided up and planted out a huge pot of dianella that was left as an anonymous gift from one of the Rewild Friday crew, knowing it's one of my favourite weaving plants. This afternoon, we've moved onto clearing and levelling a space where we want to build a deck for my bell tent near the western fence line. I've been nattering away about my visit with Gio as we work, and when I look up I see that Phillip has dug out almost a quarter of the marked square already. Things happen fast around Phillip — that is, unless we're wandering.

'Come on, time for a break,' he says, as if reading my mind. I happily down tools. We consolidate the fire and step over the broken fence on the eastern boundary, heading upstream. The prickly-moses shrubs have blossomed sunrise yellow, as have the blackwood acacias. We walk silently next to each other, our slow footfalls allowing for the internal gear shift. These wanders have become an unspoken bookend to our day. I squat down to examine a feather.

'Manna gum, I think,' Phillip says, putting a gum nut in my palm. He stands up, and I can see he's spied something. I read his body language and follow him up the hill and inside a gate to where a pile of sticks have been arranged like dried pineapple rings on top of each other. It's striking in its simple beauty. Phillip places another stick on the pile.

I allow my imagination to meet him in the invitation, and wander where the roots may lead. My mental map of the area has been expanding and fleshing out. I've confirmed that one of the blackbirds sleeps in the acacia next to my sit spot, and his mate sings to him from up at the house.

We wander past a patch of blackberry still fruiting and lean precariously over the canes. 'Yummmm,' Phillip grins at me as his long arms pluck the juiciest fruit from the middle of the thicket, where the birds and foxes haven't yet trawled.

He pauses to point down near his feet. I immediately see what he's found, an animal track running east–west, heading towards a heavy thicket. We silently signal to each other our who question and follow the track to where a narrow burrow disappears into the earth.

'Bunny?' Phillip asks.

'Maybe, but I haven't seen any sign of them.'

Phillip puts his hand at the mouth of the burrow and extends his fingers. It more than covers the entrance.

'Too small for a fox?' I say.

Phillip shrugs and picks up a tuft of fur. 'Hmmmm,' he says, raising one eyebrow.

I smile in pleasure at the shared mystery. I'm so glad we're doing this. I almost cancelled our Monday because I was feeling too busy, but now I can't contemplate missing out on the day. A compatible wandering buddy is a rare find and therefore I've mostly wandered alone. Others seem to talk a lot on their way *to* somewhere, missing the treasures and questions that put spring in the wanderer's step. Wandering is the fastest way for me to quieten my mind — a dynamic conversation between body, instincts, imagination, and the land. It's been an unexpected joy to find someone else with this sensibility.

I love having things brought to my attention that I might otherwise overlook: the shape of a branch or new growth on the casuarina. Phillip will point out plants and I'll show him birds. He's highly tuned to certain things: weather before it arrives, what nutrients a seedling is missing, the best place to plant a sapling. I think it's partly born of having impaired hearing. Without this orientation, Phillip has amplified his other senses, as his eagle eye attests. And it's partly his upbringing in the bush, running around wild on the land as part of a five-kid family, like I did. Sometimes, just the synergy of us walking together incites me to see more as I pre-empt his observations and questions, the healthy competitive streak a great goad to my awareness. Not infrequently, we gesture towards something simultaneously, as if there's a sensory porousness between us.

We reach the river. Morning-glory vine blankets the native riparian trees. Perfect basket-weaving material. I expect we're going to keep heading upstream, but Phillip pauses and points his nose uphill as if on the trail of something.

'Let's head north,' Phillip says, his gaze still held in that direction.

'North?' I question with almost a snort of surprise. 'There's nothing there but suburbs.' Phillip raises his eyebrows as if to put the question back on me. I think of Gio and catch myself in a blinkered judgement.

'North it is,' I say.

Ascending from the river, we dart between cars to traverse the main road and slip inside the first side street. We stride definitively northwards, slowing down to our usual stroll when enough distance is between us and the traffic.

'Hey, look, in the salvia,' Phillip says, and I look over to catch the silhouette of a New Holland honeyeater disappearing inside the inner skirt of the thick shrub. 'Gawd, that purple is stunning,' he says, speaking out loud the internal praise I'd already been lavishing on the velvety purple spires of flowers. A red salvia overhangs the picket fence, and Phillip picks a flower, pinches off the base, and brings it to his mouth.

'Really, that sweet?' I say, surprised, but find when I try it that a good squirt of sweet nectar shoots onto my tongue. We go back for seconds.

I scan the front yard and pattern match the English cottage's garden plants with names, uses, and memories. The yard's thick with colour and bees, despite the cool breeze. The yard next-door is laid out with keyhole gardens of winter-crop seedlings, as if the neighbours have struck a companion planting deal. A white cabbage moth flutters close to the kale, and it launches us into a conversation about effective control.

'Hey, check this out,' Phillip says and points to the ground. A string of ants are making a one-way pilgrimage along the side of the footpath. We squat down to look closer. 'They're carrying eggs,' he says, his big second finger pointing non-discriminately to the ant highway, many of whom I now notice are carrying white sacs on their backs.

'Oh yeah, I wonder if that's the rain coming?' I say. We track them down the footpath until they diverge off to the left and start ascending the trunk of a large eucalypt.

'Wow, that'd be the largest one I've seen in this neighbourhood,' Phillip says, staring up into the branches, where the ants are still climbing. 'Sixty or seventy years at least.'

The lilly-pilly trees in the street are fruiting, and we gather a handful and spit out the pips as we go. Grapefruits, pomegranates, and lemons are fruiting in front yards. The last of the tomatoes hang red and ripe in a sunny corner.

Our conversation is interrupted when a flock of silvereyes sweep through in front of us. I wonder if it could be the same mob I saw in the privet this morning. Their combined song always stops me mid-sentence, the haunting notes carrying far further than I would expect for so small a creature. The song sinks into my chest, and I close my eyes ever so briefly. When I wander, there's usually a transition point after about twenty minutes when I find myself properly arriving in my senses and I can meet the place not through gaps in the hazy filter of my thoughts but directly. It heralds a new availability to where I find myself, the start of a real conversation. I can feel that transition now, as if everything is standing out a little more in technicolour fullness, sounds sounding closer and more distinct.

The wind changes direction, and we both catch it. In another ten minutes, fat droplets start falling. Phillip turns his face to the sky and pokes out his tongue. A fat drops splashes onto it, and he bursts out laughing. They start coming faster, and I too tip my head to greet the southerly. It replies with a fat drop in my eye, making me exclaim in surprise.

Aha, this is 'weather joy' again, I think, remembering a conversation a few days earlier in my kitchen with my friend Pippa as a storm front rolled in. 'Come on,' I said. Initially resistant, once outside Pippa lifted her face to the thick raindrops that had started falling and began laughing, really laughing. 'It's weather joy,' she said with eyes open wide. 'It's probably been two years since I've really received weather like this, like a guest. If I was at home right now, I'd probably be on Netflix.'

'Yep, Nature Flix is even better!' I said, laughing with her. I reflected later on how weather joy is a particular kind of happiness threatened by our addiction to indoor comfort. I've been actively welcoming it in. Last week, the national news announced there was going to be a once-in-a-hundred-year storm and recommended Melbournians cancel all plans and batten down the hatches. Excitement built, neighbours made plans with each other over the fence, shops ran out of candles. Newspapers sporting front-page weather hype sold out across town. I awaited the storm at my sit spot wrapped in waterproof clothing. And waited. Finally, there was a downpour. And then, the sun came out. It was a storm in a teacup. All that the media could drum up was a picture of a young mum wading up to her knees in a carpark while her kids climbed down the handrail. Photos of puddles on the ground titled 'danger' were soon circulating on Facebook. There was an air of disappointment. You promised us a storm! The sentiment was almost one of betrayal and spoke to me of a yearning for something bigger than us to interrupt our neatly made plans, to bring us to jaw-dropping awe at the force of the elements, to remind us of our human vulnerability. As much as we like to think we have cocooned ourselves successfully from the bite of the elements, there is a deadening in this too.

This shower barely dampens my hair. We're not ready to go home yet. Phillip stops abruptly, and I pull up beside him. A tall shrub I would have passed by is coming out in pale-yellow rosette flowers. He brings one to his nose.

'Awww, get a whiff of this!' he says. 'What is it?'

'Hmmm, kinda cinnamon spicy,' I say. 'I have no idea, it's unfamiliar to me.' I rub the leaves between my fingers. They're rough like sandpaper. Phillip whips out his phone and posts in a Facebook plant-identification group.

'It's that easy these days,' he says, smiling broadly.

A bushy vacant block appears on our left, and we head in.

'Look at this!' I say, excited to find myself at the edge of a ring of fungus, delicate indigo-blue lace skirts and orange caps.

'Yeah, wow!' Phillip says, meeting me in my excitement. 'And check this out,' he says, directing me to a dainty creeping grevillea in flower. I join him in clearing away the weeds from its base. Caretaking is just such a natural impulse for him. It's about the fourth time he's paused on this walk to free a plant from overgrowth or adjust something to make it more beautiful. It's his love language for the world.

My caretaking includes use. I've already got pockets bulging with figs, lilly pillies, and lemons; kurrajong seeds for roasting; a stash of tinder from the palm-tree trunk; and some dianella reeds for weaving. It's a tending and appreciating of a different kind.

In *Braiding Sweetgrass*, Potawatomi grandmother and science professor Robin Wall Kimmerer describes how sweetgrass meadows that were used traditionally for basketry and ceremony doubled in vigour and density when harvested according to traditional guidelines. The meadows nobody was harvesting did worse. Sweetgrass needed the basket makers' disturbance to help them reproduce; restoring a relationship restored the wellbeing of the plant. A world untouched is a world unloved.

'North not the badlands after all?' Phillip says with a wink.

'Ha, yeah, it's quite the wander-land.'

The rain is getting heavier, and we pick up the pace, crossing the threshold of the main road, this time under the cover of dusk, like two limber deer bounding across the tarmac, and head inside for tea.

Approaching it with the sensibilities that I usually reserve for 'nature wanders', I'm starting to see a wildness in the burbs different to that of the parklands, present specifically because of its human and non-human overlays — the unplanned chaotic mishmash of indigenous and introduced, edible and non-edible plants; the places where the cultivated rubs up against the uncultivated. There's a synergy in these edge places that intrigues me with its possibilities. What lives under the open decking? What edible weed is flourishing where the water pipe leaks? What weaving vine is taking advantage of the open sunny verge? What relationship exists between the nesting currawong and its human neighbour? What food is there for a wily old fox?

The concept of wilderness is not a particularly natural idea, nor is it necessarily doing nature or us many favours. Despite the fact, of course, that these places were occupied by the First Nations and therefore never 'untouched', wilderness implies a dualistic separation between nature-land and human-land. In *The Trouble with Wilderness*, William Cronon says, 'If we allow ourselves to believe that nature, to be true, must also be wild, then our very presence in nature represents its fall.' By this thinking, the place where we are is the place where nature is not. Setting humanity and nature at conceptually opposite poles is not a good strategy for healing the rift that created such distinctions in the first place. We need to see ourselves *as* nature, rather than an enemy of nature — fall in love with our own wild natures as much as the wild nature in the mountains and valleys.

According to environmental writer Emma Marris, by limiting our definition of nature we're essentially 'stealing nature from our children'.

'The Nature Conservancy did a survey of young people, and they asked them, how often do you spend time outdoors? And only two out of five spent time outdoors at least once a week. The other three out of five were just staying inside. And when they asked them why, what are the barriers to going outside, the response of 61 per cent was, "There are no natural areas near my home."'

And well one might think that if nature is synonymous with national parks. Instead of wilderness, Marris suggests a conception of nature that is more inclusive, such as places with the presence of multiple species, places where life is thriving — like the little vacant lot that Phillip and I found. I could have lingered way longer in just that quarter acre. It was an experience of being immersed by completely autonomous, self-willed nature. These untended parts of our city, city edge, and town existence, Marris says, are arguably more wild than a carefully managed national park. 'It takes a lot of work to make those places look untouched,' she says. 'And they're expensive to get to. They're hard to visit.'

Marris is advocating for recognition of the value of what she calls 'rambunctious gardens' — a hybrid between wild nature and human management — places right under our noses, she says, that we don't even notice. 'If we dismiss these new natures as not acceptable or trashy or no good, we might as well just pave them over.'

Marris points to the example of an abandoned elevated railway in the heart of Philadelphia that has turned into a wild meadow through natural seeding. Biosurveys have found over fifty plant species. Rather than calling such places 'weedy', scientists have started referring to them as 'novel ecosystems', because of their unusual mix of native and non-native species.

Novel ecosystems in Australian cities support almost 380 nationally listed threatened species, substantially more than all other non-urban areas on a unit-area basis. In fact, 30 per cent of threatened species were found to occur in cities — the powerful owl that I've been hearing call up the river amongst them.

Of course, much has to do with the degree of tree cover. It came as a surprise to me when I met with Ian Shears, the primary urban forester at Melbourne City Council, that Melbourne is actually classified as a forest. London too just scrapes in under the UN definition of any landscape with over 20 per cent tree canopy.

'They don't make forests like they used to,' I joked to Ian as we ordered lunch in a noisy arcade cafe. He laughed but wouldn't concede.

'We're aiming for 40 per cent canopy cover by 2040 within the city,' he said. 'That really is an urban forest.'

I thought about it for a few moments before responding. 'So, really, you could propose that the city is primarily a forest with buildings in it, rather than a built environment that you add trees to.'

'You could, yes,' he said, munching down on a panini. 'Doesn't that change things?'

It was an enjoyable intellectual flip in perspective, but not one I seriously entertained. I know a forest. And Melbourne ain't it. London's got to be even further from it. Still, the ratio of green to grey makes a huge difference to the experience of nature. And Ian seemed to be a guy with a whole lot of power to make that happen.

'We often think of the trees as the lungs of our city, but they are also, in some ways, our heart and soul,' he said.

Ian Shears told me how people 'erupted' after the council planted a wildflower meadow on top of a stormwater-harvesting system. 'I had more positive feedback from that than any horticultural intervention because there was a degree of wildness about it. It wasn't petunias or bedding begonias. It activated the part of us that still lives on the savannah.'

Some years ago, the city gave each of Melbourne's seventy thousand urban trees an ID number and email address so people could report in on their condition or on problems related to vandalism or hazards. Instead, what they received was thousands of expressions of loving fan mail, existential questions, and witty tree banter.

... today I was struck, not by a branch, but by your radiant beauty. You must get messages like this all the time. You're such an attractive tree, read the letter to a green leaf elm.

Babe. I am sorry that you're so sick. Can I climb you one last time? Strip down that bark for me baby, it'll make you feel better.

Hi Tree 1022794, How's it going? I walk past you each day at uni, it's really great to see you out in the sun now that the scaffolding is down around Building 100. Hope it all goes well with the photosynthesis. All the best.

Dear Nettle, I just moved in three months ago and I'm very glad that I can talk to you through this system. I live in the first floor and I can actually see you through my window! I'm having trouble sleeping at night because of the noise of cars and ambulances at night, hope you're not suffering that much and be able to have a good sleep. Thank you for blocking the noises from the street and wish the birds don't do harm to you. Pleasant to meet you and have a nice day! Cheers!

Dear Rose Gum, Over the past year I have cycled by you each day and want you to know how much joy you give me. No matter the weather or what is happening around you, you are strong, elegant and beautiful. I wanted you to know. Love.

'It gave people permission to communicate with trees on a very intimate level,' Ian said. 'If I'd asked permission of Council for that project, they would have said no. It surprised us all and just went viral.'

Meanwhile, stories from the Rewild Friday crew have started trickling in. Michelle has started sitting in her backyard.

Tas has settled on a spot right on Merri Creek, five minutes' walk from his home. Directly in front of him is a tree that has fallen across the creek and dammed a pile of rubbish. Yet within a couple of sits, Tas has noticed a scrub-wren nest within the branches of the fallen tree, and it's popular with willie wagtails and grey fantails too. 'Despite the disgusting mess, there's wildlife everywhere,' he tells me. 'Now even when I don't see many birds, I always come away excited about one thing that happened. Even something small.'

And I can't stop smiling when I receive this email from Elena:

At my sit spot this morning at dawn, the raindrops fall so gently they create a layer of tiny water bubbles on the shawl my Mum knitted for me and now warmly envelops me.

I greet the tree I sit beside almost every day. We share the view, seeing and being seen, and being a part of, not separate. The resident magpies, this is their place I come to visit. Their morning song fills my heart with joy, full to almost bursting. The river flows, ever so slightly higher than the day before, orange fungi pop up seemingly overnight. Where are the ravens and lorikeets this morning? I notice these things now.

The additional plastic bottles and other discarded rubbish floating in the river showing the latest thoughtlessness being washed down to the sea. I arrive at work, into a grey office with grey walls and grey carpet. I greet my warm-

hearted colleagues. They mention it's miserable out there this morning. As my thoughts return to the morning sit spot I find the words ... 'all weather is beautiful' pops out of my mouth before I can filter it. Oops! They all stop and look at me now. Then one of my colleagues, an older woman with a warm and fierce heart comes over to me and kisses my hand, she says 'I love you Elena'. I'm asked if I've added marijuana to my breakfast this morning to be talking this way about the weather on this wintery day. I laugh with them, at the incongruence of what I said; recognising absurdity, but also total sense, as I realise it's the sit spot; the tree, the magpies, the flowing river, they've come with me to work and beyond.

Elena, it seems, has claimed her sit spot — slowing down, paying attention to the trembling mysteries that Martin Shaw speaks of — and in turn, her sit spot is claiming her.

Chapter 4

Friday afternoon. I'm on my bike. There's a storm heading this way. A distant flash of lightning. Thunder growls like a large cat in the eastern hills. My brakes screech to a halt. Oh god, I'm finally home. I drag the bike up the kerb, unlatch the gate, and throw it onto the rack. *Stop. I just have to stop.* Doesn't look like there's anyone else home at the Riverhouse. All they would get is a grunt anyway.

It's power-down time. My iPhone nestles in the palm of my hand. It looks so innocuous, so innocent. But it's a sticky-webbed imposter. With a hint of revenge, I press my finger a little harder than necessary on the side button. It resists for a few seconds. Petulant thing. And then finally dims to black. I open the top drawer of my desk, throw it in, and slam the drawer shut. Gone. Banished. My body issues a long involuntary sigh, as if released from the grip of a large hand that's been squeezing me tighter and tighter. I need to hide out where no one can find me, somewhere free of human agendas and future plans. I pull on my jacket, grab a blanket, and head out the back door.

The wind picks up my hair and blows it around my face. I button my jacket up to my neck, unwilling to be discouraged. The wind slaps at the trees in violent gusts. Yes, I'm pissed off too. *Off with the world's head.* Each slow step downhill is a statement of rebellion. I refuse to rush anymore. Refuse to be this city's call girl. Down down down, I'm going down, and I ain't coming back again.

A small willy-willy of sand kicks up around my legs as I cross the fire circle. I kick back at it and stride through defiantly, collapse in the hammock, and wrap the blanket around myself like a cocoon.

Alone, alone at long last. Thunder moans. I put my hands over my heart, a bid to make contact with this person who's become a stranger. *Remember me?* The sudden stillness amplifies the crackling in my head. There's too many tabs open, the software is wigging out. The system needs a serious defrag. I breathe long and deep.

I peek out over the blanket through a hole in the hammock at my sit spot. The grass has started growing back where I usually sit. It looks forlorn. I've hardly been down here the entire last month. The last entry in my nature journal is an excited series of questions about the fantail cuckoo that I had been hearing and just caught a glimpse of. Since then, I got swept up in a maelstrom of busyness and have completely dropped the thread. And now the lead has gone stone cold. So much I've missed! Major seasonal transition markers. So much for an everyday sit spot. The longer the time lapse between visits, the harder it is to return, like an unreturned phone call from a friend that gets more difficult to bridge with every day of silence that passes. Do the birds and trees feel my absence too? I feel guilty for ghosting them. The thunder peels out across a darkening sky, and the grey shrike-thrush pings out a single bell-like chime from above. I wrap the blanket tighter around me. Oh god, how did I let myself get this fricking overwhelmed. I feel like I've just been spat out the other side of a cyclone.

I don't even know when it started. There was a moment a few weeks ago when I looked at my diary and realised that I was facilitating groups every weekend for the next month in addition to having evening plans most nights. I went through the classic stages of grief — denial (telling myself it wasn't so bad), tantruming at myself, bargaining about what I could get out of. In the end, I gave up and let the juggernaut of life roll right over me.

There has been no end to the human doings: the emails, appointments, phone calls, organising, promoting, listening, teaching, correspondence, communication, meetings, tasks. It hasn't all been work, either. Parties, celebrations, dinners, catch-ups. It's not like I've been whittling away hours on Netflix; all the activity has been in service of the simple things in life — meaningful engagement with the world, nourishing friendships, exchanges of ideas and resources. It's just that the web of human relationships is anything but simple. My heart feels like a revolving door. I'm angry at myself for not recognising the wall I was about to hit before my nervous system began crackling like the lightning on the horizon and sent the lights out. My tolerance used to be way higher. I just can't push through like I once did.

I'm conscious of the absence of my phone. *Sign of addiction*, I tell myself. I'm so used to it tugging at my wrist like a needy child. There's life before and life after a smartphone entered my life. I remember the day I bought it: I walked out of the shop bemused about my new toy. Yet pretty soon the toy soon felt like it was playing me. Sometimes, I find myself picking it up without a particular purpose and succumbing to the promise of something more exciting than the current moment awaiting me within its vast reach. The act triggers an instant reward of dopamine — the same stuff that's released in the brain of a rat when it beats a maze and gets the cheese. It soon becomes less about the cheese and more about avoiding the withdrawal symptoms of not getting the dopamine hit. It doesn't feel like I'm in a maze so much as on a hamster wheel, chasing the next hit on social media.

And my phone addiction is comparatively mild. I switch it off overnight, enjoy flight mode often when I'm not on planes, and have the dubious reputation of rarely answering. Yesterday, a client told me how first thing in the morning and last thing at night she scrolls through the Facebook feed on her phone. I felt nauseous at the thought. The image arose of a creek choked with weed. An ex-housemate came back from Bali inspired by her device-free time and didn't put a SIM card in her phone for weeks, and then only with limited apps, and started leaving it at home when she went out. Three months later, she was texting while in yin-yoga poses.

Humans created five billion gigabytes of digital information in 2003; in 2013, it took only ten minutes to produce the same amount of data. No wonder there's an exhaustion epidemic, with us having to chew through the overwhelming amount of information we feed our systems every day. The hammock seems to sag with the extra weight of my tiredness. A friend talks of his fatigue as if it's a shameful secret. At work you wouldn't be able to tell, he says, as he remains upbeat and self-medicates with coffee. 'But if I could, I'd just crawl into a ball and sleep for days.' Last Friday, Elena was looking drawn and pinched. 'Yeah,' she said. 'It's like my body is getting increasingly unwilling to play along anymore. It's rebelling.' Philosopher and social commentator Charles Eisenstein has a theory that increasingly common ailments like chronic fatigue and depression are the body's quiet mutiny against the specific demands placed upon us in our hyper-modern world. 'When our soul-body is saying No to life,' Eisenstein says, 'the first thing to ask is, "Is life as I am living it the right life for me right now?" When the soul-body is saying No to participation in the world, the first thing to ask is, "Does the world as it is presented me merit my full participation?"' It's no fun anymore. We're collectively packing up our toys and going home.

I packed off to the bush for a year as part of my own rebellion against this colonisation of my mind and time — and that was before I even had a smartphone. I craved endless acres of unstructured time. And had a growing curiosity to see what might emerge within my psyche when unbound by schedules, goals, obligations, or expectations. I went cold turkey, ditching my old mobile phone, even hiding my single watch under the bed. The freezing-cold splash of creek water on my face every morning reminded me that there was nowhere I had to be and nothing I had to do. It was what I needed to shock me out of the enculturated mind maps in which I equated my value with how much I could fit into a day. And so I learnt to sit, to wait, to watch, to know without knowing how I knew, to wander off trail for as long as my feet cared. To rest, move, eat, and play when I wanted rather than when a clock declared; to create when inspired, focus when required, and at other times rest deeply in the dreamy singsong of the wooded symphony.

As I gradually slowed down, the life in and outside of me grew full with a kind of all-pervasive presence. The spacious rooms of the forest were anything but vacant, moments sometimes ripening to such intensity that I would burst out laughing or crying or both at the heartbreaking beauty of the unknowable mystery contained in all things.

In this hedged-in world of calendars and clocks and micro-second immediacy, mystery can hide unnoticed in the shadows, cracks, and crevices of the psyche. I made a commitment when I moved to the city. *Okay*, I told myself, *we're moving to the city, and if you start speeding up, then you'll also start forgetting. So we're not going to do that. Deal?*

The deal is clearly off as I lie here twitching and fizzing out on the city crack of overstimulation, driven to distraction by my own limbic system. This is what I feared, that the never-enoughness would be impossible to avoid in the cosmopolitan hype, the distractions too alluring, the slope too slippery. And I've slid all the way down to the bottom again.

'Busyness is not a virtue, it's a choice,' a friend reprimanded me when I took forever to return her email. Yet accepting responsibility makes busyness all the more painful. I'm not a victim to the city's rush hour. On a treadmill, there's always a big red stop button. But so often, I don't make that choice. I just hang on for the ride until the machine bucks me off.

I re-centre my hands over my heart, hoping their warmth will fill the hollow inside. It's ironic that when I overfill life is when I feel most empty. I become a stranger to myself and a stranger to the world.

Thunder rolls another log across the roof of the sky. For a moment, one bushel in the top of the canopy takes on the silhouette of a woman's head. Her face tips towards the amassing clouds, her eyes closed. It's an ephemeral sculpture, a gust of wind quickly disassembling it. That sound! How might I begin to explain the sound the wind makes through gum leaves? I toy with words: sanguine tinsel, the chinking of beetle wings. Could each species of eucalypt have its own distinct sound? The question intrigues me. My heart warms as if a little ember has just blown to light in the same breeze. A raven flies in to land in the branches above and caws, its beard fanning out below its beak. I burrow down into my blanket and drift off to sleep while the storm passes over.

From the park bench I'm seated on, I watch kids throw breadcrumbs to a gaggle of fat white geese, which attack the crumbs with familiar aggression. Behind them is Studley Boathouse, a quaint English-style tearoom where you can hire unwieldy boats that never go far, the occupants satisfied with taking selfies as they paddle around in circles. The moment I spy Sean walking down the hill, I realise that this is the place he brought me for a walk some time ago when I'd not long arrived in Melbourne. It's funny how the roles are reversed and now I'm the local welcoming him back.

He walks towards me in a checked shirt tucked into black jeans and heavy lace-up boots.

'Hello, Ms Dunn,' he says, walking straight into my hug. We grip tightly, breathing each other in. He smells the same. Apart from a few more grey strands in his goatee, there are no obvious changes. He leans his head on one side just as I remember, touching his hand to his driver's cap as if in greeting, his blue eyes smiling warmly. He still looks like he could equally emerge from an Irish pub with a book of poetry under his arm or from a woolshed.

'Welcome back,' I say, as he takes my hand.

'I guess it is a return.' He laughs a little nervously. 'Maybe a comeback.' We stand in silent contemplation of each other, reading the stories in each other's faces.

I first met Sean at the Melbourne book launch of *My Year Without Matches*. He was the MC on the cold September evening that a couple of hundred people turned up to sit around a fire for a conversation between myself and my friend Maya, another nature writer. Sean and I met up for coffee a couple of days later to continue a private conversation. The second meeting confirmed the strange experience that being with Sean was like being with myself in another body and gender. Our instant attraction felt like kinship. It was uncanny.

At the time, I was the wolf come in from the woods trying to navigate the city streets with paws for feet, and him the street-smart one who had learnt how to tuck away the grey fur under black coats and heavy-soled boots. He couldn't hide it from me, though, tufts of canine hair escaping from his every word and gesture. We outed each other in the most delightful way. It's unnerving to be so seen, or rather to see oneself in another so alike. Sean tells me that just being around me puts his nose back on the trail any time he loses the scent. In my early days with Ostii, I felt like a wild seed blown in from a great wind, trying to find purchase; around Sean, the wild seed found ground.

Then a year ago, Sean decided to live on a farm with his two sons. We messaged a few times, but he was mostly offline. Now we drop hands and turn shoulder to shoulder along the same river path we walked some years ago.

'Look at you, living riverside,' he says.

'Yeah, and swimming almost daily!' I say.

'Good to hear you haven't lost your edge,' Sean says with a sideways smile.

'So ... back from the woods,' I say, opening the gate for any stories that might want to be caught. I know the impossibility of trying to sum up even a little of what happens when one's life is so dramatically reshaped; it feels intrusive even for me to knock at the door of these experiences.

'Yep, been a couple of months now,' he says buoyantly and then registers my enquiring glance. 'Well, what to say, you know how it is, Dunn.' His answer intimates that the experience was indeed as rich and beautiful as I had imagined, and that he's still trying to find his footing back here.

'You've kinda missed the just-hatched me,' he says, 'when I was looking at city life through a glass pane, seeing all its beauty and ugliness. Now I'm mostly just *in* it.'

'That quick, hey? Well, what's the view like from the other side of the pane, looking back at the bush you?' I ask.

'Well ...' Sean clears his throat. 'I guess I'm seeing how deeply my sense of self had expanded, beyond this sack of skin, to include, and be informed by, my surroundings. No longer just a mind carried around in a body.' He laughs. 'I feel like I'm preaching to the converted.'

'Pretend you're explaining it at your book launch,' I half-joke, knowing that in a few weeks I'm going to be the one helping Sean launch the book of poetry he wrote while out there.

Sean squats at a verge by the river and pulls up some grass, which he rolls in his fingers.

'It's like the default egoic human state is separation — I'm here, the world's out there. But what I came to experience more and more was a deep state of belonging, like a cessation of that existential separation. It wasn't something I pursued. I didn't arrive in that place driven by an idea. It wasn't something I could argue with. I guess it was a feeling of deep rest. Of being-with. And that state is really profound, for so many reasons. Beyond feeling nice, I came to recognise it as a place from which I wouldn't commit any acts of violence.'

'Violence as in actions born of not-enoughness?' I say.

'Exactly. It became my default state out there. I'd come back to the city occasionally for a weekend, and the first few times I felt like I was being psychically shredded. Arriving big and expanded or porous in some way, and by the end of the weekend I'd be feeling discombobulated and exhausted. Lying in my bed out there on the land, my sense of self was about two hundred metres across, in an area mostly empty of humans. Translate that to North Carlton and I had semitrailers, traffic, and people moving through myself! So, by necessity, I had to learn to consciously draw in my energy body before entering the city.'

I was reminded of the discernment required in knowing when to dial down the senses last week when I received an email from a woman who had dropped in for a single Rewild Friday. She'd said that after a day of practising sensory expansion, on the train ride home she became quite nauseous and realised she was soaking in too much city debris. In less dense information zones, one can really spread out. I entertained the image of Sean being rolled by a semitrailer in the middle of the night as his expanded bush self spread like lava through the streets.

'I felt like out on the land my thoughts and busyness of mind slowed down to a point where I wouldn't be thinking much at all,' says Sean. 'My body and mind became quiet and receptive. Once that happens, there's a way in which what does come to mind is of a better quality.'

'And now?'

'Hmm. What I see so clearly now is how urban existence — coupled with media advertising and popular culture — tends towards self-importance and self-absorption in a really unhealthy way. In the city, I so often feel like I need to do something to justify my existence. There's something like a low-level anxiety … What am I going to be today? What am I going to generate? I really noticed that, coming back.'

'Same here. I associate that feeling with the hum of rubber on bitumen. People all going somewhere to do something.'

'Coming back, I found it really hard to wake up to the day and feel at ease simply knowing that life is happening around me, doing its thing, and that that is enough. I reckon it has something to do with how we relate to time. By contrast, living on the land, I found I had a healthy and emergent sense of my place in the scheme of things.'

I've also been contemplating my relationship with time since chatting to one of my mentors — master tracker, naturalist, and bird-language expert Jon Young. I was telling him of my frustration with lack of time for my sit spot, and in his usual tangential way of teaching, he shared a story with me from his recent adventure with the Naro Bushmen of the Kalahari Desert. Noticing one of the elders' dislike of wristwatches, Jon enquired about this through his translator. The Bushman gave a short sentence pockmarked with consonant clicks that was quickly translated as 'whenever someone looks at a watch, the next thing they do is rude'. Rude! I laughed out loud. What a trenchant way of describing the quality of social transactions that take place within compressed schedules and narrow bandwidths of attention. Jon went on to describe the long-winded greeting customs of the Kalahari. When gathering, there might be a full forty-five-minute period in which greetings take place, involving asking after your family, the land of your birth, the weather, any stories from the day before, and so on, and including many warm handshakes. And then the same the next day. Time is spacious, cyclical, generous.

After hearing this story, I watched more closely for signs of rudeness in myself. It was there in subtle ways: winding up a conversation too quickly, not calling my mum back, not separating the soft plastics properly, turning the tap off too harshly so that the washer wears. Maybe what we're doing to the planet is a macro-version of this rudeness. Because in the moment of rushing, one has already broken off connection and is operating within a state of separation, the place from which violence originates.

The sun comes out, and Sean and I sprawl out on the grass. I roll over on one side and prop myself up on my left elbow, eager to hear more.

'Sounds like you're saying we just have to suck it up in the city and live from a place of disconnection,' I say.

Sean draws out a stem of grass and puts it between his teeth. 'When I got back, I tried to do all the same things I loved in the bush, like "I'll go for walks every day", but it's all felt like a lame imitation. I found myself getting pretty down, again feeling like life was getting away from me. Like I'm offering nothing and creating nothing. I don't want to live like that, so I decided to try and find a bridge to tap into that experience of deep participation more readily.'

Sean and I have talked previously about free-form dance being a way that we both find access to that authentic wild expression in the city, but I have a feeling he's talking about something more fundamental.

'I experimented with focusing on simple moments of connection and letting them expand, like scratching my son's back while watching *Star Wars*, or noticing the clouds shift shape.'

'So, more closely attending?' I say in a half-prompt.

'Yes. The bridge itself, not something I have to do or go looking for. The bridge is presence itself. Living out bush, it's easy to focus outwards for the source of inspiration, but here I need to focus my attention back inwards and in the everyday.'

Interesting. It's similar to the invitation Jon gave me after hearing the conclusion I'd come to that living in a contracted state of awareness is a fait accompli of city life. He was fierce in his refusal of this assumption.

'Creation mind is available wherever you are,' Jon said, the phrase borrowed from his Bushman friends to describe the state of mind they drop into when out tracking and hunting. It's akin to the 'wild mind' state that Sean and I talk of, characterised by a present-moment focus, sensory expansion, and full-bodied engagement with the world — a kind of deep participation in the inner and outer landscape simultaneously. I argued with Jon about the impossibility of accessing creation mind while multitasking in the city, but he was stubborn.

'Creation mind exists in every moment, regardless of what you're doing or not doing. You're never going to not be busy. Find the spaciousness within the moment itself. Widen your lens of attention, and you'll find it.'

I glance down at the river to see a darter with its wings outstretched and drying. It appears completely at ease in creation mind. Sean reads my thoughts.

'See, our greatest teacher is right there,' Sean says, and we pause to absorb the image. I stretch out my arms and let the sun soak into my wings. Some block of grief thaws out and catches me by surprise.

'I'm kinda sick of the struggle, Sean. It's the same tussle between being and doing that I've been dancing with all these years, the feeling of failing if I lose the place of beauty and grace.'

Sean chews philosophically. 'Well, if it sounds like I've worked it all out … I haven't,' he says. 'After a year in the woods, I look around and I'm more aware of the way the modern mind, with its self-actualisation obsession, is quite oblivious to grace. We don't realise that we're in a dance with an incredible dancing partner. So we push and push, as though we have to make everything happen by force of will. Living so far from the rhythms of nature allows us the delusion of control. But there's a wisdom in the woods. You quickly learn that *to be* is *to be in relationship*. That state I was talking about, of deep presence … you can't force that. You can't choose communion with nature, because you're only half of the conversation. Just like you can't put yourself to sleep, you can only put yourself to bed. Then you wait and sleep comes for you. I guess what I'm saying is, I've learnt the importance of making myself available through presence. Then I wait, trusting. Soon enough, she comes for me, and we dance.'

'So it's more about the intention itself to receive grace when she comes for you,' I say.

'Yes. But you know, both ways of being are essential to human experience — the experience of separation *and* the experience of belonging. It's majestic really, this movement between deep participation and separation, wild mind and city mind. I reckon it's what characterises our species.'

Sean's pocket starts ringing, and he pulls out a dinosaur of a phone. I burst out laughing.

'What's that you got there, grandpa?'

'Oh yeah, for my fortieth birthday I bought myself a dumb phone — a Nokia with eleven different functions. Eleven! It was hard to find one, and bloody difficult to transfer my contacts and get it working.'

'Wow, that's a strong statement,' I say, trying to imagine life now without my smartphone.

'Our lives are made up of the days, hours, and minutes, right? And how many minutes do I absent-mindedly check my phone messages, or the weather, or *The Guardian*? At the time, it can seem like, well, there's nothing wrong with checking *The Guardian*, but cumulatively those minutes make up my days.'

For a few minutes, I entertain the thought of firing up my old Nokia, until I remember Google Maps and proclaim profusely that I couldn't do without the tracking device.

'You did once not so long ago!'

'Yes, and I was constantly lost!'

'I print maps off the way we used to,' he laughs. 'And meet interesting people while asking for directions.'

'And is your mind wilder for it?' I poke.

'Well, yes, actually,' Sean says. 'When I moved back into the city, I could see how much of life here is set up around the goal of avoiding ourselves in the empty moments — the constant bells and whistles of my phone. On the tram, instead of looking at my phone, I look out the window and see the rusting paint on the side of a building, or a crooked tree. It's like I see myself reflected. And it might fill me with an existential loneliness, but there's something healthy in that.'

I look back out to the darter on the log. It's true, sometimes it's in those empty spaces where I feel a real conversation begins.

'Surely it's not the tool, but how we use it,' I argue, still trying to imagine life without Google Maps. 'My bird app is awesome.'

In *The Nature Principle*, Richard Louv talks of the potential synergy that exists between nature and technology if they are in right relation. 'The ultimate multitasking is to live simultaneously in both the digital and physical worlds, using computers to maximise our powers to process intellectual data and natural environments to ignite our senses and accelerate our ability to learn and feel — combining the resurfaced "primitive" powers of our ancestors with the digital speed of our teenagers.' If my sit spot had a reminder in my Google Calendar, I would not so easily miss my appointment with it.

'As Thoreau once said, Dunn, men have become the tools of their tools,' says Sean, throwing a ball of grass in my face.

We laugh and then fall into silent contemplation, watching the water flow down the river.

The conversations with Jon and Sean work on me over the next few days. How we spend our minutes is how we spend our lives, the micro-choices we make in each moment adding up to a lifetime. Move towards that which connects you and away from what disconnects you, advised Jon. I dance between creation mind and separation.

It's a cool night, one in which I could easily snuggle up inside, but instead I pull on a black puffer jacket and black jeans, ready to court a wild mind under the cover of darkness. Out on the main road, a pedestrian crossing squeals permission to walk, and I obey, tucking my hands deep inside the feathered down of my pockets. Flying foxes are launching themselves like little furry rockets out into the suburbs in search of fruit, and I follow in the wake of their northern trajectory.

What if accessing creation mind is truly not about where I am? Maybe it's possible to inhabit the same flow state as the tracker running after the springbok while pacing the pavement. The passing cars are wild cats with shining eyes. What if instead of an obstacle, the pressures of the urban environment could generate a creative urgency towards creation mind? What if there's a new possibility for how it might show up for us?

I take a small road off to the left. It's wide and edged on both sides with large plane trees, their roots buckling the asphalt. I walk in the middle of the road, where above the leaves reach for each other from opposite sides. I have a sudden impulse to lie down. The road is cold and hard under my back. Apart from the white of my face, I blur into the bitumen. I search the sky above for the few visible stars. The leaves tessellate a snowflake-like pattern that appears to be expanding. The knowing that I may need to jump up at the sound of a car heightens my sense of aliveness. For a few moments, the stars, road, trees, houses, traffic, and myself mingle and dance into a single experience of pure beingness. I rest on the tarmac, feeling the acres of earth under me and the limitless sky above. A great relief washes through me. It may not be the creation mind of our evolutionary ancestors on the African savannahs, but it's a version of it, lying here on this gritty street in the dark as the bats fly over.

Busyness is an extrovert. It will happily occupy all the space in the room. Filling up life to over capacity is a sure-fire way to ensure that nothing surprising or numinous happens to us, our boat anchored so tightly to the shore that we can't possibly discover the hidden golden reef just beyond our awareness. Maybe Jon can drop into creation mind while answering emails, but I suspect that for most of us it requires at least a pause in proceedings to allow us to hear its soft rapping on the door.

Creation mind is quiet. It leans up against tree trunks and whispers through portals of clouds and winds and between the sentences in conversation. It requires the attending to another, more secretive and subtle, life behind and beyond the charismatic and the diarised. This secret life is conducted in small gestures. Words of love to the new moon. A tuned ear to the first bubbles of the kettle. The delicious swish of the first toe in a warm bath. Songs sung as the storm rolls in. A long gaze into the eyes of another. Skyward and slyward glances at the play of light on leaves. Crunching underfoot the last frost of winter.

The secret life is found the moment I register the lilt of the scarlet honeyeaters cavorting in the gum blossoms. It's the private dark humour I share with the wood ducks on the jetty, the irrational hope I see within the green shoot of the single squash seed that my thumb pressed into the earth some weeks ago. Together, we shiver when the south wind blows in, and together we spread our wings on the log in the shallows when the sun shines. It's the sway of my hips as I ride the tram in peak time, the compassion extended to the grief cry on the face of a stranger on the street, the hug that is simultaneously an embrace from the arms of the world. It's an equal division of attention to the seen and unseen, the explicit and the implicit, matter and mystery.

Joseph Campbell names the tending of this secret life an 'absolute necessity': 'You must have a room, or a certain hour or so a day, where you don't know what was in the newspapers that morning, you don't know who your friends are, you don't know what you owe anybody, you don't know what anybody owes to you. This is a place where you can simply experience and bring forth what you are and what you might be. This is the place of creative incubation. At first you may find that nothing happens there. But if you have a sacred place and use it, something eventually will happen.'

My sit spot is that place. Leaving the middle world of street level, the descent to the riverbank is a signal to my soul that I'm ready to listen, that it's the underworld to which I now turn my attention. Down there, in the fecund damp, as I lean against my tree, things pad in that otherwise keep to the shadows: feelings and knowings, ideas and curiosities. They are daunted by the bright lights of attention and scared off by agendas. They could be small things — what needs to be planted in my garden; someone I must call — or larger movements of the heart. This time strengthens my pause muscle.

Elena has developed a ritual to mark the start of a more conversational nature at her sit spot, by offering her backrest tree some of her tea. One day, she told me, she was at her sit spot feeling distracted when 'all of a sudden, a leaf fell down, and I looked at it, and it hit me. I was at my last sip, and the tree was like, "Hey, I'm here." I offered it the last drops.'

Committing to her sit spot, Elena told me last Friday about an experience that had blown her away.

'I was sitting in wide-angle vision, and all of a sudden everything flipped, and I realised that instead of being looking, I was being looked at by consciousness itself. It isn't a one-way thing, is it?'

Over his fortieth birthday weekend, Sean lit a backyard fire that burnt for forty-eight hours, and welcomed friends and family to drop in to share a story or two. The ashes from that fire live now on Sean's mantelpiece as a way to remind himself of the presence of his mortality. And using the old Nokia is a simple discipline to help him keep located in the fleeting moments of life.

These little knots I tie myself in again and again, perhaps they are necessary binds so that I may remember afresh the unbound. And yet rushing is also a product of love. There is so much in this life that I care for and that cares for me that I just want to run around and smother the earth in kisses. There is no end to the length of string of things to do, and what trails behind without complaint is the sweet earth, the quiet voice of the soul. It's a matter of making myself available to these moments. To hold the list of things to do not as a yoke, but as an invitation to love the world more fully.

I will join Sean in this dance between deep participation and separation, this movement from one side of the bridge to the other, and see if I can find a place of acceptance somewhere in the middle.

Chapter 5

I squat next to a patch of grass and slowly reach my hand down to separate one of the younger leaves of a dandelion plant from the rosette. It's going to be quite bitter given it's flowering. Still, this is not a consideration the large Channel 9 camera pointed in my face cares for. It's just hungry for the climactic moment. I briefly hold the leaf outside my mouth before chewing down on it. It is bitter and slightly astringent, but I give none of this away in my expression.

As she moves in with the microphone, the TV reporter balances precariously near the pavement edge in her black stilettos. 'So why would you eat a weed rather than go to the shops for lettuce?'

A couple of dog walkers pass by, and I catch their sympathetic smile. Being in front of a national news camera is not an enviable place for most people.

'Well,' I say, wondering if I have greens stuck in my teeth, 'one person's weed is another person's food or medicine. Many of the plants we call weeds have been eaten as a nutritional vegetable for thousands of years. In many countries, some are still staples today. And, of course, it's free.'

'So tell us about what you're eating,' the reporter says, her bright red lips distracting me.

'I just ate a leaf from the dandelion plant, which most people know from the seed heads you blow to make a wish. The roots and flowers are edible too. In fact, the United States Department of Agriculture Bulletin ranked dandelions in the top four green vegetables in terms of overall nutritional value. It has the richest source of betacarotene of any green vegetable, is a liver cleanser, purifies the blood and lowers blood pressure, amongst other things.'

'And how does it taste?' the reporter baits.

'Great!' I say, smiling innocently.

The news coverage is due to a media release that went out this week about the Great Local Lunch, an initiative of the annual Sustainable Living Festival whereby chefs create a four-course meal for 250 ticketed folk solely from 'crowd-farmed' urban-backyard produce, as well as from weeds that I will facilitate a group to forage. The project is about promoting sustainable urban agriculture — edible gardens not as a hobby but as an increasingly vital source of food for the city. The project draws attention to the fact that a 2016 Food Future Report found that Melbourne will need to produce at least 60 per cent more food to provide for its predicted seven million residents by 2050. This city needs to become farm as well as dinner table.

In the evening, I watch the story from a TV-owning friend's living room. 'Here you are!' she yells when I finally appear. It's a short and mostly pictorial infotainment segment at the end of the news, with a single sound bite from me.

'Is it just me or is it bizarre that eating a plant that has been consumed for thousands of years is of national significance?' I ask rhetorically, watching the last scene, where I chow down on the leaf. Still, none in my teeth.

'It's not so much the what as the how,' she says. 'It's the foraging bit.'

The next morning, I wake to a call from *The Age* national newspaper, requesting a photo shoot and interview. They've seen the Channel 9 story and want a bite on the proverbial weed too. I'm beginning to feel like the flavour of the month. The following day, Min and I sit on the front couch with a cup of tea exclaiming over the photo of me amidst a sea of pink amaranth flowers that covers three-quarters of a broadsheet page. The piece is titled 'Weeds Make Good Grub — Apparently'.

'Seriously?' I say, pinning the paper down on the bricks as a breeze tries to lift it. 'Page five of the national newspaper. Is it that exciting?'

Min stirs her tea and ponders. 'Well, for me it taps into the fascination with the idea of stepping outside of the monetary economy. You know, that the wild self can have a place in the city.'

'There's power in self-reliance, for sure,' I say. 'It's like one of the city's secret fantasies.'

With everything dished up to us so effortlessly — water, food, heat, and light — my feeling is that foraging is an act that taps into the yearning to feel less like a passive consumer and more like an active agent. Even with the cameras on, those moments of plucking a plant from the earth for food heralded instant satisfaction. At the opposite end of the spectrum, I recently opened the freezer at home to find a packet of mixed berries from the USA. I couldn't even stomach the thought of putting something so location-specific into my smoothie. A red-ripe summer berry belongs to a particular bush in a particular place on a sunny day after just the right amount of rain. I can taste it now. How much can a food be removed from its source — its season, locality, and grower — and still call itself the same food? Some part of me needs to forget what food is in order to be satisfied with ingesting a far-flung berry, both for its sake and my own.

I got more serious about foraging when wild foods became a necessary part of my daily diet in the bush. Some plants took me completely by surprise, like the spikey little shrub that I had hardly given a thought to that bordered the path to my shelter — which then revealed itself to be the maker of sour little fruits high in vitamin C. Even when it was in homeopathic amounts like in geebung berries, eating from the land was a core practice of belonging. It attuned and endeared me to my environment in a similar way that tithing any of my survival needs from the land did. Through need.

These days, down the road at the appropriately named Oasis wholefoods store, a magic carpet can fly me from an Istanbul bazaar for halva to an Indian market for curry paste and then straight to a Viennese chocolate shop for candy. The village market has gone global; food is another commodity, a fetish, a plaything.

The more direct my relationship with what sustains me, the more direct my experience of connection. I wonder how much I could actually grow or glean here? Not just plants, but carbohydrates and protein? Could I actually feed myself if I needed to from the city's wild larder? It's time to get my research hat on.

A weed eater named Grubb, one of Melbourne's famous foragers, bikes decisively towards where I wait on the edge of the parklands. When I first got a copy of Adam Grubb's co-authored book *The Weed Forager's Handbook*, I marvelled at yet another name and vocation correlation. His handbook is ubiquitous on Melbourne bookshelves, designed to fit neatly in the back pocket of skinny jeans. 'Weeds are the ultimate convenience food. They ask of you no money, no search for a park in the supermarket, no planting, no watering,' reads the first line. But this forager is clearly more into his greens than his grubs. He's probably also into a good nettle cider, I judge as he alights with beard and black skinny jeans, clearly an inner-north native. Adam has been one of the pioneers taking weed eating out of the grassroots and into five-hat restaurants and hipster cafes, leading tours of the food-laden cobblestone laneways and parks of the city and modelling it as a smart and sassy way to dine.

As he ties up his bike, I tell him about the media frenzy of the past few days. 'I thought you'd already made foraging trendy?' I joke. Adam laughs in condolence.

'There's still a whole lot of grass-eating hurdles,' he says, reminding me it's still for the most part a back-pocket activity. 'To make the leap from hunting the wild asparagus in the supermarket to foraging in a vacant lot can initially just feel so foreign. Most folks still don't associate non-cultivated spaces with food. People are still more comfortable buying the $5 spinach from the supermarket rather than the more nutritious option growing freely in their backyard,' he says, bending down to pick a stem of purslane.

I suggested this meeting so I can check out what's available in readiness for the Great Local Lunch workshop this coming weekend. We wander down by the creek, nibbling and comparing notes as we go, finding mallow, milk thistle, cleavers, clover, oxalis, onion weed, cat's ear, dandelion, wild lettuce, wild radish, chickweed, plantain, and dock. We lock horns very politely on one plant with an arrow-shaped leaf. I'm sure it's one of the goosefoots in the *Chenopodium* family; Adam is equally sure it's a native lookalike. We whip out our phones to cross-check, like duelling weed geeks. I concede defeat. Adam notices my excitement to add a new plant to my database.

'The whole foraging project is highly enjoyable,' he says, reaching down for a particularly tasty morsel of plantain. 'In fact, a lot of cheap and free stuff you can do in the city is more enjoyable than the stuff we spend money on and that actually makes life less enjoyable in the long run. Like a houseful of random possessions that seemed wonderful when you bought them but now seems to demand more care, organising, and storage space than you have capacity for.'

Grubb's broader philosophy, shared by co-author Annie Raser-Rowland in *The Art of Frugal Hedonism*, reframes sidestepping of the monetary economy as inherently pleasurable rather than a sacrifice. At the end of our wander, we pick through the various wild brassicas for a bag of greens, and when I get home, add them to the pot that's heating up last night's curry for lunch. As I munch, I reflect anew on the fact that most of these wild edibles are actually more nutritious than the vegetable greens growing in my garden, easier to grow, abundant, free, and pretty tasty. Still, how many times do I walk over them without picking a handful for dinner? Despite everything, I'm just so conditioned that food is this and not that. And conditioned to think that food procurement is something to turn my attention to at mealtime (by opening the fridge), rather than something that is opportunistic and continuous.

The night before the foraging workshop, I get a text from the producer of the event. *There's a long waiting list. Expect some queue jumpers. Good luck!* In the morning, there are at least twenty keen beans, including a few kids, at the meeting place, with bags and gloves. I do my best to mark off names but don't have the heart to turn away those eager enough to line up on a Saturday morning for a weed lesson. This knowledge should be free anyway, I tell myself.

My co-facilitator, Taj Scicluna, appears, looking true to her other name as the Perma Pixie, with baskets and bottles in hand. A veritable encyclopedia on all things weedy and wild, Taj has bleached dreadlocks that spill out in long snakes from a knot on the crown of her head, and her tanned chest sports a broad tattoo of the hexagonal cells of a beehive. A dandelion flower is botanically lifelike along one arm.

'So many people. Wow!' she says, spilling over with enthusiasm, and some of the participants look at her as if she's just sprouted from the soil. Taj pulls out a couple of bottles of dandelion tonic from her basket and proceeds to send them around the crowd with instructions to place a few drops under the tongue.

'It's like Popeye food,' she jokes. It's great to be in collaboration again — the two of us ran a series of sacred ecology weekends some years back.

We give the initial disclaimers and warnings on pesticide sprays and stormwater run-off before I hand over to Taj.

'Okay, so who's ever eaten a foraged weed before?' she asks the crowd, scanning for a response. Half a dozen hands go up.

'Does nettle tea count?' a woman asks.

'Yes, if it was wild-harvested,' Taj replies. The woman nods, satisfied.

'What's the one that the chickens love?' a guy from the back asks, and I crane my head to see him.

'Chickweed?' I guess.

'That was it. My girlfriend put it in a smoothie for me once,' he says.

'Any other experiences?' I prompt. Nothing forthcoming. 'Okay, let's have one right now. Look down.' Everyone's head drops. 'You might have some sticky seeds on your jeans care of a plant I knew as sticky beak when I was growing up, or farmer's friend.' There are murmurs of recognition. 'The green leaves of this plant are a daily vegetable in some parts of Africa. Have a taste for yourself,' I say, and I pick a few sprigs from the plants nearby and send them around the circle. It's a good icebreaker: neighbours sharing leaves with each other and giggling at the novelty of putting something from beneath their feet into their mouth.

'There, now you've all eaten a weed!' Taj says.

The surprised delight on the faces of the crowd reminds me of my own weedy aha moment, and I spontaneously decide to share the story. In my early twenties, I went to stay with some old forest-activist-back-to-the-landers, Marg and Barry. The place was like nothing I had seen before, an owner-built, cave-like hobbit home in thick forest on the edge of the creek. One evening, I accompanied Marg to the veggie patch out the back to pick greens for dinner and was surprised to find that instead of heading for the furrowed cultivation zone, she took her basket to the edges where weeds tall as my head grew. 'Fat hen,' she said, plucking the grey-green arrow-shaped leaves and putting them in her basket. We steamed a basket-full and stuffed them into a pie crust along with pine nuts and fetta, served with homemade tomato relish. 'Just like spinach,' Barry concurred.

Once the acquaintance was made, I began picking the fresh tips of fat hen whenever I came across it. It became my first wild edible-plant friend. I later found out that seeds of fat hen have been found within the stomachs of several mummified peat-bog men as well as at many other sites of ancient civilisation.

I can see everyone listening and engaged. Stories are always the best teaching tool.

'I began taking notice of other weeds and then suddenly got fascinated and bought all the field guides and for a little while got obsessed. Now when I look at the ground, it's no longer homogenous lawn but rich with meaning and value — my foraging brain by default scanning for pattern matches as I recognise edible and medicinal leaves as well as connections to people and places.'

'Right now, all I see is green,' one woman says, a bit despondent.

'That's going to change really soon,' I say. We split up into two groups and I lead one mob down towards a pair of ancient mulberries in a field where I know the weeds find a favourable micro-climate.

The woman's sentiment is not uncommon. Sydney celebrity forager Diego Bonetto calls it 'plant blindness' and points out the number of branded products we would recognise compared to names of common plants in our area.

'If you gain confidence with one plant, it will open the door to others. Here's a good one to start with,' I say, squatting down to point out the narrow-leafed plant growing from a rosette. 'Plantain — known as white man's footprint because it's followed settlement all over the world.'

I pick the seed head and hold it up. 'This is where psyllium comes from.'

'Really!' exclaims a young woman in two brown plaits who immediately kneels down to find her own. 'Super expensive in the shops.'

'The leaves are edible, and if you chew them up they make a quick but powerful poultice to draw out infection.'

With their curiosity piqued, I introduce them to other 'rock star' weeds — dandelion, mallow, wild brassica, and dock — so named because, like others such as amaranth, mustard greens, clover, lambsquarters, and chickweed, these powerbrokers of the weed world grow everywhere people live (from the hottest deserts to the Arctic Circle), are easily identifiable, and can be used as both food and medicines with minimal preparation.

We stumble across purslane up on the slope, and I can't bring myself to pass it by — one of my personal faves. 'It's rich in omega-3 fatty acids,' I say, feeling like a tour guide. An older woman with a dark-haired high ponytail jostles to the front so she's right next to me and listens intently, her eyes veritably glistening as I talk about how purslane is still used in some Mediterranean countries. I watch her bend down and pick a single small fleshy leaf and bring it to her mouth as a wine taster might, smelling it and taking a tiny nibble before putting the entire leaf in her mouth. She smiles as she swallows. 'Wow, that's delicious. I think I remember my grandmother picking this. She called it glistrada.'

'I reckon your memory is probably correct,' I smile. 'This might be your doorway plant.'

To everyone, I say, 'Okay, weed dating over, it's time to forage, we've got a crowd to feed.' I let them loose with a bunch of bags on the creek verge. It's thick with wild brassica, dock, mallow, and plantain. The foragers turn towards the real teachers growing at their feet. Some of them take a few quiet moments by the creek before beginning to forage. Some others I can see are being diligent in selecting just a few of the young leaves, following my advice on taking only a certain amount from each plant.

'Even weeds?' one guy had asked.

'Let's think of them as wild edibles,' I encouraged him.

I was confronted by my own unconscious plant hierarchy not long ago. My herbalist friend Rosie came over to play, and we collected bags of sticky cleavers that we mashed up in apple-cider vinegar, mixed with violet, and left to steep as an acetum — a salad dressing and medicinal lymphatic support. I really wanted Rosie to help me try my hand at making mallow milk and had earmarked the single large mallow in the backyard; I had my digging stick poised when Rosie said, 'Do you really want to take this one?' I paused and looked at her. 'Or do you want to let it seed other mallows?' *But it's just a weed*, I heard part of me respond, instantly recognising my conditioned response. Me and my stick backed away.

The foragers have spread out up and down the creek. The woman with the ponytail is ambling along with a dreamy smile on her face. Someone starts whistling. I sit down in a patch of sun to take a breather before joining them. It's a bucolic scene, the sun filtering through the leaves onto the burbling creek, slow-moving figures bent over with their hands and minds engaged in the simple task of finding food by the bank. The way they collectively move reminds me of mountain goats inching along from one plant to the next, absorbed in the task. There's a settledness that I suspect is partly because I invited them to put their phones in flight mode.

No one is fiddling around with technology. Everyone is here. It's rare. And profoundly restful.

I spy a patch of mallow nearby, and it motivates me to move. The first few leaves tumble into an empty cloth bag at the pinch of finger and thumb. Pretty soon, I drop into forager's flow, eyes scanning for familiar shapes as hands pluck and pick and dig. My fingers work deftly, softly encouraging out the odd wild onion root, plucking off the young shoots. I come across a blackberry nightshade ripe with dark-purple berries and squeeze the sweet juice onto my tongue. In the periphery of my vision, I remain aware of the larger group, like a shepherd. We constellate around each other in slowly shifting patterns.

A warm joy spreads through my whole being. I know this feeling: foraging joy. It just seems so natural, a task I feel entirely *designed* for. And I am. For at least the last 1.8 million years of our six-to-seven-million-year progression from ape to human, foraging was our way of life — our colour vision is said to have evolved as a direct consequence of the need to identify coloured fruit in the forests. Even as most of our ancestors became agriculturists over the last ten thousand years, wild foods remained an important part of the diet until recently. Judging from the pure joy it seems to elicit, I imagine it must connect to some pleasure centre of the brain that remembers and rewards this food-finding activity.

The Rewild Friday crew have voted foraging days as their favourite so far. Recently, we went on an excursion out of town to harvest edible mushrooms in the pine forests. Tas returned from his long wander with an entirely different presence about him, the furrow in his forehead gone.

'Something switched on in me,' he told me around the lunch fire. 'It was the combination of wandering and foraging and all the sensory opening we've been doing. It all just came alive.' Words weren't necessary. I could see it in his eyes.

The pleasure of foraging derives from the fulfillment of what I've heard First Nations people call our Original Instructions — the human design fundamentals that allow us to live well with each other and with the land. Other traditions might speak of the dharma, the tao, or natural law. In *Braiding Sweetgrass*, Robin Wall Kimmerer explains the concept of Original Instructions through the Skywoman creation story, which encodes within it ethical prescriptions for respectful hunting, family life, and ceremonies. Rather than instructions or commandments, says Kimmerer, Original Instructions are 'like a compass: they provide an orientation but not a map. The work of living is creating that map for yourself. How to follow the Original Instructions will be different for each of us and different for every era.'

Original Instructions reside within Traditional Ecological Knowledge (TEK), which includes complex knowledge of how to balance biological and human cultures. According to Kimmerer, foraging is an opportunity to practise and learn this covenant of reciprocity, and she offers a list of her own principles of 'honourable harvest': *Ask permission of the ones whose lives you seek. Abide by the answer. Never take the first. Never take the last. Harvest in a way that minimises harm. Take only what you need, and leave some for others. Use everything that you take. Take only that which is given to you. Share it, as the earth has shared with you. Be grateful. Reciprocate the gift. Sustain the ones who sustain you, and the earth will last forever.*

One of the kids runs up to me with a leaf to double-check identification. I give him the thumbs up, and he beams, running back to the plant to pluck a handful of the youngest leaves before moving on. Foraging engenders a sense of responsibility, initially for correct identification and then in the act of an honourable harvest. There is a bright pride in his face that comes with being given this responsibility. It's a small taste of engaging in a TEK practice, being trusted to step into that reciprocal relationship with the land that is both caretaker and apprentice.

When approached as a sacred skill, foraging becomes a kind of communion with plant and place, an embodied ritual of experiencing life as directly supported by the particular elements, biota, and features of a landscape. There are no intermediaries, no warning labels, no long-haul flights, sprays, or shelf-life preservatives required. It's as immediate as hand-to-mouth. Inherent in this ritual is the reminder that ultimately the earth is the hand that feeds us, and our part of the exchange is exactly as Kimmerer suggests — by way of gratitude, generosity, and care.

I was offered a warning from the wild plants when I broke the creed. I was in a rush while out scouting for suitable foraging locations along the creek for an upcoming Rewild Friday. I approached what I assumed in my quick assessment was wild fennel, plucked some seeds, tossed them in my mouth, bit down, and froze. No aniseed flavour. The adrenaline kicked in microseconds before I spat them out. It was poison hemlock, the plant that killed Socrates. There's not much room for mindless munching when wild eating. If you want to rely on your wits, then first open your eyes, the plant told me. And if you want to teach this sacred relationship, then take the responsibility seriously. Start with the principle 'don't die today'. Got it.

I cooee the crew to gather, and they stroll in from the fields, smiling and laughing. We meet up the top with Taj's group, who look equally cheerful. The morning's forage amounts to half a dozen good-sized bags of greens, which will be put on ice until delivered to the Great Local Lunch chefs tomorrow. Taj and I invite a debrief. The initial chitchat about speed dating weeds opens the way for deeper reflection.

'I really noticed how, unlike what I buy, I'm not in control of this food. I take only what I'm given, in a way, and there's a humility in that,' says one woman from Taj's group, adjusting her red neck scarf as she speaks.

'Yeah,' pipes up her friend next to her as an idea lands. 'It's really wild food, isn't it? It hasn't been bred for taste or size. No one owns it. It's not reliant on anyone to feed or water it. It's resilient.' I wait as she thinks more about what she's just said. 'Wow, I've never really considered the difference between wild and cultivated food in that way. How cool!'

'Yep, and like we talked about with the honourable harvest, it exists in a gift economy rather than the monetary economy, and that creates a different relationship to the food too. An equality perhaps,' I suggest.

'Yes,' continues the woman excitedly, 'I don't feel like I'm consuming so much as … I don't know … living!'

Taj and I farewell the participants, many of whom give us hugs of appreciation and show us their party bag of take-home weeds. I head home excited to try out a new dock-seed recipe. While the group foraged, I had waded into the middle of the creek, where there was an island of large dock plants, and filled a brown paper bag with the seed heads. Most of it is chaff — the papery seed sheath — but I throw it all into a coffee grinder to make a gritty meal, to which I add mashed sweet potato, purslane leaves, a cup of sprouted buckwheat, half a cup of almond meal, a couple of beaten eggs, and some herbs, salt, and pepper. Shaping it into patties, I fry them in some oil and set them out on plates with rocket, avocado, grated carrot, and a sprig of purslane for garnish. On the side, I make a wild weedy salad, and a pot of millet that I stir through with turmeric, plantain, amaranth, purslane, fat hen, dock, dandelion, and a few mallow flowers.

It's a full plate already, but then I remember the olives that Phillip and I harvested from an Italian friend's grandmother's backyard. We filled buckets, including enough for nanna, and then sat them in brine to leach out the bitterness. In one afternoon, we easily gathered a year's supply. I pop one in my mouth. It's softer than the ones in the store, and the pip's slightly larger, but still — salty and delicious and as local as you can get.

I roll out the Persian rug on the deck, lay out the lunch, and call Min and Megan from their rooms. I wish Phillip was here to share it too, but he moved out recently to his own bungalow not far away, and has been replaced by Tully, an avid gardener and guitarist.

'Wow, what are we eating?' Megan says settling onto a cushion on the edge of the rug.

'Weeds mostly,' I reply as Min joins us.

We join hands for a brief food blessing before digging in.

'Mmmm. Love the patties,' Min says.

There is joy in sharing this food. No fossil fuels were burnt in refrigerated trucking, no water drained from fragile river systems, no fertiliser allowed to run off into those same waterways, no insects poisoned by pesticide-coated leaves. I have full faith in this food. It's a free feast and a feast for the free. My next step is to learn more about the bush food in my area, and I know just the man to call on.

'There are three S's to consider in bush food,' says Doug, leaning somewhat awkwardly on the park fence. He adjusts his thick glasses and looks at me expectantly. I have no idea what he's about to say. I look over to where my friend Elizabeth is standing, hands on hips and notebook in hand. She shrugs my way.

'Staples, snacks, and survival foods,' says Doug. Elizabeth starts scribbling. 'For instance, what would this be?' he asks, swivelling around and pointing through the fence towards the clump of what I know well as *Lomandra longifolia*.

'Snack food?' I say, knowing how little carbohydrate I have ever managed to garner from the fleshy base of the leaves.

'Yeah, probably. Maybe sometimes a survival food,' says Doug, leaning over the fence to pull up a few from the inner circle.

Elizabeth takes one and mirrors me as I pull on the end of the reed with my front teeth like a rabbit. She giggles. 'Wouldn't want to survive on this,' she says. 'Is this saltbush berry?' she asks, picking a peppercorn-sized ruby-red fruit from a low-growing grey shrub on the side of the river trail.

'Sure is — there are four types of saltbush along here, and all the berries are edible,' Doug says, smiling.

'Oh, is that what that is!' I squat to examine the plant, which I'd suspected was edible. Albeit small, the berries are abundant, and I pick a handful and toss them in my mouth, where they burst into sweet flavour. To think I've been passing these by every day. I take a picture with my phone, and add a corresponding number and name in my notebook.

Another twenty minutes and one hundred metres later, my notebook is overflowing with new knowledge. I met Doug a couple of years ago, when he participated in a Vision Quest program I was guiding, and I quickly discovered his encyclopaedic knowledge of local bush foods. While fairly well versed in bush foods from Sydney to the Queensland border, I'm finding the drier woodlands of Victoria shyer in revealing themselves. I associate green and verdant with edible, so brown and grassy equals hungry in my mind.

There are berry snacks — dianella, pigface, cherry ballart, tree violet, and myoporum. Drinks from the sweet bursaria and banksia flowers, and flavours like the drooping cassinia, a close relative of which (common cassinia) can be used as a curry-leaf substitute. Even the casuarina offers a snack in the form of the rather dry flower bud, which is edible when young, while the leaves offer a thirst suppressant. Elizabeth and I lick the leaves of the manna gum free of the sweet lerps that fall off like sugar crystals.

'What about survival foods?' I ask.

'It's all around us,' Doug says cryptically, waving his arms emphatically and waiting for us to guess. 'River red gums,' he says with emphasis. 'The seeds are high in nutrients and were crushed up and made into cakes in hard times. I've also heard stories that the inner bark was also eaten. But you'd really only want to go there if you were pretty bloody hungry.'

Walking on, Doug picks a couple of small leaves from a vine growing over the fence and hands them to us. I know it as clematis, and I sometimes harvest the fluffy seeds heads for tinder bundles. Doug picks another and holds it out in front of him as if he's about to perform a magic trick. We watch in suspense as he crushes it between his fingers, stuffs it up inside one nostril, blocks the other, and inhales with a sharp sniff, followed by a quick head shake and nose rub.

'Your turn,' he says in a voice slightly higher in pitch. Elizabeth and I give each other a quizzical look.

Within seconds, the entire right side of my nose is on wasabi-like fire, which then mellows into a general giggly light-headedness.

'Woo, bush snuff!' I say, giggling. 'The fourth S.'

'Yeah, kinda,' Doug says, '*Clematis microphylla* was once used as a natural anaesthetic and headache cure, and also the roots were pounded up as a poultice for aching joints, although if left on for more than a few minutes can cause blisters.'

Sedatives seem to be the fourth S in Doug's knowledge bank, which seems to tend towards unusual and little-promulgated Indigenous plant use. I learn that hop bush can be chewed as an anaesthetic (as long as you don't swallow it), blackwood bark can cure rheumatism and stun fish, sweet pittosporum gum relieves ant bites, and the ever-flowering goodenia was used to put children to sleep by rubbing the leaves on their fingers, which they then licked.

Already with this download from Doug, I feel more at home on this riverbank, and will do more so when I ground this knowledge into practice — stopping on my walks to snack and snuff. There are hierarchies of intimacy in any relationship. Being able to recognise and name a species is a start: you've got their number. Next might be getting familiar with their seasonal moods: their flowering, fruiting, bark-shedding, colouring patterns. Maybe then you might recognise a plant in the dark by its shape or feel. But when you understand its gifts in the way of food, medicine, and fibre and begin to relate to the plant in this way, then a real intimacy starts to be forged.

'What about staples? What plant food might I eat around here regularly?'

'Well, that's a sad story,' says Doug, poking a stick in the direction of the fat clumps of wallaby and spear grass that dominate the ground cover. 'These gaps between tussocks is where you would have found lilies and tubers such as murnong daisy. Ethnobotanist Dr Beth Gott from Monash University found that a murnong tuber has nearly ten times the nutrient properties of a standard white potato. The original people farmed the land with murnong, planting the top back in the soil so that it grew abundantly. But because of grazing, they have all but disappeared, which of course had calamitous results for the people who depended upon them. There is a theory that Indigenous people switched to cat's ear,' Doug says, pointing out the prostrate dandelion-looking weed on the path verge, 'but it would have been a poor substitute if a substitute at all.'

Not far away, a riparian patch of Merri Creek is being repopulated with native tubers, celebrated with an annual harvest that has become something of a festival. I was amongst a couple of hundred others who a few months ago witnessed digging sticks unearth thumb-sized tubers of murnong and chocolate daisy.

Aunty Bev sat in the middle of the natural amphitheatre alongside two other white-haired Wurundjeri women, talking to us about the process as the tubers baked on the coals behind them. 'Most people lived off murnong,' she stated simply. 'People need to come into an area and have something in the ground they can eat.' After three hours of cooking, snippets of murnong were handed around on a platter of paperbark. I held the inch-long piece in my mouth for some time, imbibing its smoky, earthy flavour. It felt like a sacrament of place.

Something in the ground to eat. How true. I scan the ground around the tussocks, imagining it busting through with lilies, bulbs, and orchids, their fragile flowers poking up through the hard ground in winter and early spring, waving their street signs of delicious food. In his book *Dark Emu*, Boonwurrung man Bruce Pascoe also points to what Doug alluded to — that the First Nations engaged in sowing, harvesting, irrigating, and storing food plants like the murnong. This not only questions the hunter-gatherer label, but also begins to blur the distinction between wild and cultivated food.

Cultivation exists on a spectrum. The wild progenitors of Indigenous-managed plant populations were minimally modified if at all. Modern fruit and vegetables have been bred over generations to favour sweeter taste, fewer seeds, larger size, uniform ripening, and the ability to travel long distances. The sacrifice is in nutrition. Wild foods are on average significantly more nutritionally dense. This protected from chronic disease as well as avoiding massive alterations to the landscape. Traditional diets also combined a mixture of cooked and uncooked foods, eaten in season with little refinement and no simple sugars. The other major difference was diversity. There are over twenty thousand species of edible plants in the world, yet 75 per cent of the world's food today is generated from only twelve plants and five animal species. Compare that to the Palaeolithic ancestors of the Hausa of West Africa, who utilised 119 food plants, or the !Kung of Africa, who ate eighty-five plants and fifty-four animals. Diversity in the diet protects against nutritional deficiencies, as different foods contain different amounts and ratios of vitamins, minerals, and essential fatty acids. It's a mix known as 'deep nutrition'.

I planted my own deep-nutrition seeds just last week, embedding a tray of the flyaway murnong in soil-raising mix with added charcoal so as to mimic the periodic burning that the murnong would have been accustomed to. Alongside the murnong, I planted native seeds of lemon myrtle, temperate macadamia, blackwood (*Acacia melanoxylon*), and a rare native parsnip that my friend Jamie, a self-taught bush-food geek, gave me. Handing me the seeds, he also wrapped my wrist with some string he'd spun from local bootlace bush, saying, 'When this snaps off, you need to go bush again.' Every day, I've been checking on whether any of the seeds have sprouted, but nothing yet.

I twist the string on my wrist as I tell Doug about my own murnong plantation, and he smiles warmly. We wind up the bush-food tour under the shade of a wattle, where we open the seed pods and guestimate how long it might take to collect enough seed to make damper. 'A long time,' was the collective answer.

I farewell Doug with effusive gratitude for so generously sharing his knowledge, and for dedicating his life to bringing back knowledge on native foods. I look down at my notebook. It's spilling over with new knowledge, and all within cooee of home.

Could I really feed myself within the city's bounds? I do a quick scan of the suburbs, mentally mapping where I know food grows — fruit trees spilling over laneways, kurrajong seeds hanging in pods, field mushrooms in the abandoned paddock by the highway, native plantations thick with kangaroo apple and saltbush, olives. There are plums and mulberries sprouted from discarded pips along creek lines, figs poking out of alleyways, wild greens thick enough to mow. And there's the cultivated: my own garden and those close by.

I was gawking over the fence at the house at the end of the street when the owner, Caz, appeared from behind a raised garden bed where she'd been busy planting kale seeds. 'It's the day after full moon — the perfect time to plant,' she said cheerfully in gloves and hat. Her garden had been the first thing I noticed on the street, my daily walk to the park necessitating a ducking under their fruiting feijoa, fig, macadamia, and carob trees while chooks cackled in the background. Caz let me in on the local self-sufficiency secrets that she and husband, Chris, have been fine-tuning during their thirty years' habitation here — collecting coffee grinds and chook scraps daily from the local cafe, loading their wheelbarrow with poo from the horse stables at the CBD's edge. She sent me home with a big bunch of parsley and an open invitation.

Then there's the house that I saw spilling over with persimmon, so I put a note in their letterbox asking if they needed help eating them. 'Come over!' was the enthusiastic text-message reply.

I've recently installed a beehive and am being mentored in natural beekeeping by my friend Emma. If I keep filling out an edible map of my environment and filling my gardens, perhaps next autumn I'll see if I can live within the means of the urban pantry entirely — eating only what I forage, glean, or grow for a period of time. And as my mentor Tom Brown Jr once told me, 'the greater the need, the greater the result'. The seed has been planted.

Chapter 6

'What's a possum?'

I look into the girl's blue eyes, which are staring intently back into mine, and struggle to hide my shock. Blonde ringlets are packed away under a large blue bow, some escaping to dance about her pale round face. She wears stockings, a red jumper dress, and shoes that look like they might never have ventured off carpet. There is no hint of humour in her face. A throng of kids around us have taken to flicking the brushtail-possum poo that I have just pointed out on the grass towards each other.

'Well ...' I begin, realising I've never had to describe to anyone what this most common of our native mammals looks like. 'It's a bit like a cat that lives up a tree, with a cute pink nose and pointy ears.'

'Oh,' says the girl, looking up into the branches of the Moreton Bay fig above us. 'Are there any in this tree?'

Quite possibly. It's one of the few trees in the park without the plastic possum-proof guard. I follow her gaze.

'Can you see it?' she says, squinting.

'Actually, it sleeps in the day and comes out at night,' I say.

'Oh,' she says with a quizzical look on her face, 'I'd like to see one,' and turns on her heels to join the others.

Wow — never seen a possum. I remember how as a child I would often give an apple to who we called 'mother possum' up on the verandah eaves where she liked to hang out, watching as she shyly ventured forward to claim the gift. Knowing her, I wasn't scared at night when she trip-trapped across the roof or hissed and hocked from the branches of the fig outside my window. I hope the girl does convince someone to take her to the park with a torch after dark and catch sight of the two round shiny green eyes staring back from the branches. It's one of the few charismatic native fauna easily visible in the city. Although perhaps her experience might be less cute and fluffy and more the chainsaw-wielding hiss of the ubiquitous urban-guerrilla possum atop the park garbage bin.

I turn around to find Mel teaching the group how to cooee. 'Okay, let's send it out to the city,' she says and faces towards the skyscrapers a block away. It's a restrained call that goes out, but even so, a couple of the kids put their hands over their mouths after hearing the faint echo that comes back.

'And again, this time bigger!' Mel says, and this time the cooo-eeee bounces back to us from the glass in stereo sound. A few pedestrians look around, confused, for a few seconds as our sound pebble lands in the noisy city pond. The kids clap in excitement.

Mel and I are here at the invitation of a large insurance firm housed in one of the city towers that overlooks the park. The childcare room in the tower recreates nature in the form of a mural, a square of artificial grass, and a fake log. The harried childcare organiser who told us that many of the kids live in city apartments was keen to get them out into some real green for the school holidays. We perched waiting for them like birds on a branch in the far corner of the park, which afforded a view of the western side of the CBD. It was peak pedestrian hour, and despite the fact we were metres away, the throng kept their heads down and funnelled towards the entrance to the underground train station. The kids appeared in two neat lines of fluoro-yellow vests marching up the hill. A few of them caught sight of us and started waving, then looked sideways at their carers as an afterthought, to see if this display of enthusiasm was allowed. 'Well, we won't be playing camouflage games,' Mel whispered before nervously jumping down from the perch to greet them. Up close, the vests were even more stark, and I was relieved when their minders started collecting them in neat piles.

'Who are the wild ones, we are the wild ones ...' Mel begins a brave song, accentuated by stamps of her feet, in an attempt to build group anticipation. I look on, a little bemused. It's certainly a different scene from her weekly homeschool nature-play group. I popped in last week, locating them by the sound of young voices in unison as they skipped over the creek footbridge, brandishing what looked like wands. It was a bit like walking into a storybook. I sometimes help out — less for the pay and more for the fun of tumbling about with some wild youngsters. A few familiar mud-streaked faces ran towards me to show me the wands that they had whittled with vegetable peelers and sanded to a smooth finish. 'Wow, it's silky as a beach stone,' I exclaimed, meeting their enthusiasm.

'Are you wild?' Mel lingers on the last line of the song, looking around and sniffing the air for the wild ones. The tiny boy next to me shakes his head furiously. 'My brother is,' says one girl, tight-lipped and with her arms crossed, as if dobbing him in for a crime. 'My cousin is wild,' says another girl accusingly. Mel and I look at each other with eyebrows raised — 'wild' clearly synonymous with 'naughty' here.

We had discussed what kinds of activities would suit this bunch, particularly in regard to risk. Kids need age-specific risks that allow them to demonstrate capability, build rapport with mentors, and test their edges. Nature has always been great at providing those maturity thresholds. If we're confident in a kid's ability, they might start whittling at age nine or even younger. As they carve their first piece of wood, I can almost see them growing upwards with pride. At a nature-play conference recently, though, two presenters talked about appropriate risk in terms of uneven ground and diverse textures. How the yardstick has changed.

We settle on the fail-safe fun of a scavenger hunt, which pretty much lets them choose their own adventure. Off they race in small groups, ferreting out feathers and treasures in hidden corners and crevices of the manicured landscape. One of the items on the list is simply labelled 'bugs'. Pretty soon, Mel and I are being tapped on the shoulder every few minutes with a new slater, worm, or otherwise-unidentified many-legged creature. We bring out the magnifying glasses, and the kids jostle to get a look at the growing number of insects that are beetling about in the Buglet Town bark enclosure we've made.

'Wow,' one kid says, looking at the red-tailed crawly on his arm. 'This is one crazy dude.'

This micro world hidden within the grass is also the charismatic fauna of the city — abundant, diverse, fascinating, each species with its own stories to tell. I've been taking more notice of the smaller things at my sit spot too — not only beetles and spiders, but also tinier leafed plants. From the macro to the micro and back again.

At closing, the possum girl approaches me. 'Thanks for learning us about proper nature,' she says somewhat formally.

'Did you have fun?' I ask.

'Yep, look what I found,' says the girl and reaches into her pocket to show me a large gumnut. 'It's a fairy house.'

'Yes, it is. Maybe you could have a look for some more around home?'

'Yes, and create a fairy village,' she says, the sentence trailing off into the far-fetched future. 'Goodbye,' she says and runs off to join the others donning fluoro vests, this time with pockets bulging with bugs.

~

'Where do you think it went? Uphill maybe?' my nine-year-old neighbour Eva says quizzically, her brow furrowed. She looks ahead, and back behind her, contemplating the unlikely migration of a large macropod into her backyard.

'That's a good question. If you were a scared kangaroo, where do you think you'd head to from here?' I ask.

Eva's bright red hair shines in the late-afternoon autumn sun, her face pale and sprinkled with fine freckles of similar colour. She looks around with renewed focus, the question I posed working on her empathically.

'I think over there,' she says, pointing upstream into a neighbour's backyard.

'Okay, let's go check it out,' I say and follow in her confident footsteps.

We're on the hunt for 'Stampy', the name new-housemate Tully has given to the eastern grey kangaroo that must have got separated from his mob upstream and ended up in our backyard about two weeks ago.

My first Stampy sighting was dramatic. I was on Zoom with Jon Young, and had spent the morning preparing questions to ask him. I thought it fitting that I take the call at my sit spot. He immediately picked up on the screech of the lorikeets and currawongs in the background, remembering them from his sole visit to Australia. We launched into a spirited discussion about city sit spots, me confessing that I had yet again fallen into a bit of a rut with my questions.

'One good thing to do is ask yourself the question "What am I missing?"'

As if on cue, in that moment I heard a loud thump and spun around to see a large male kangaroo jump out of the bushes as if being chased. I too jumped up to standing position, the phone sliding from my hand and into the grass just shy of the river embankment. I barely registered Jon's voice still chatting away as the roo bolted not metres past me into the neighbouring yard, where he proceeded to crash into the boat and then ricochet into the fence. Could a roo really make it past all the suburban obstacles to get this deep into the city?

'Oh my god!' I exclaimed. I watched the distressed animal bounce around the yard, unsuccessfully trying to head downstream. 'Claire, are you there?' I heard faintly. My brain hurt with the cognitive dissonance: a large distressed mammal in my urban backyard, a very important phone call.

'A kangaroo ... it's here ... no way, I can't believe it ... oh god, it's crashing into the fence again ...' I gave the bemused Jon a real-time commentary.

'What were you saying about nothing happening at your sit spot? Remember, there's always more going on at your sit spot than you think. Don't assume anything!' he laughed with childlike joy.

Stampy's presence has thrown a much-needed wonder bomb into my sit-spot routine. I haven't seen him again, although I have found fresh scats and one of his lying-down spots where the grass has been flattened above the jetty. He's definitely still here. I had bumped into Eva and her mother on the street and invited Eva to help me track Stampy. Having rarely seen her down by the river, I was surprised and delighted when both agreed, mostly for Eva's sake, but also because kids are sharp detectives.

Eva walks straight over to some fresh scat and squats down next to it.

'Do you think these could be Stampy's?' I ask, squatting next to her. Eva screws up her nose and pokes at them gingerly with a stick.

'What has it been eating?' I ask.

She pokes the deposit open. 'That looks like grass. Yes, probably yes, they are, I would say definitely yes, they are Stampy's. Maybe that means we're getting close,' she whispers, looking behind her.

'We'll have to creep.'

'Good idea,' I say quietly and crouch to meet her level, the two of us stalking like thieves into the neighbour's yard. I smile at her audible commentary.

'There's long grass up there, you'd like that ... come on, Stampy, we won't hurt you ... we just want to say hello ... look here, more poo!'

I ask some more questions to get her thinking about the age of the scat. She grows more excited. 'We're definitely on the trail.'

Eva's curiosity is contagious. I can feel myself being pulled into child mind, future thoughts flying away in the immediacy of the mystery. Where is this animal? Can it see or sense us? Is it scared? Lonely? I get down on my knees to examine the ground more closely. This kangaroo could be a Tasmanian tiger, given the degree of our thrill in the tracking, yet not ten kilometres away we could walk past a mob of fifty of them and nod a hello in passing: value and interest are relative to rarity. And in this landscape, a roo is near extinct, which makes for an intoxicating scent trail.

An image of finding Stampy hiding beneath some blackberry bushes flashes into my mind. It's the direction Eva is heading. Kids are such natural trackers. It's a science, but it's also an art, relying as much on logic and knowledge as the imagination and instinct that Eva is overflowing with right now. Together, we make a good team. I can feel Stampy's heavy weight at the other end of the rope we're tugging on.

We follow our noses and the odd scat until the grass gives way to dirt on another property boundary. I can't see any visible tracks, but point Eva to a possible claw mark in the ground. She starts scanning the ground with her fingers. 'There's one,' she says quickly, jumping on a small indentation, which motivates a new and earnest movement onwards. I follow in her wake, recalling the remainder of my conversation with Jon.

He was even more animated after Stampy's interruption, launching into a spirited soliloquy on what I soon realised was a summation of his decades of research across continents and ancient cultures into the technology of nature connection.

'We have a nervous system. We have a circulatory system. But what most people don't realise is that we also have a connectivity system.'

I had turned around to sit facing the direction that Stampy went, in the hope of another glimpse as I listened.

'Connection is a direct conversation between our nervous system and nature itself. We are all hardwired to make relationships with trees, animals, landforms. If you had ample unstructured time in nature as a child, then your connectivity system was naturally activated — you'd be curious, engaged, questioning, imaginative, excited — the qualities of a good tracker.'

I reflected on the fact that while I'd had a fair amount of unstructured playtime in nature as a kid, there was still a sense of something missing.

'Mentoring,' said Jon. 'That's what ancient cultures knew how to do. Teaching came in the form of ceremony, games, and stories. Learning came from the very real need to hunt and gather a wide variety of edible and medicinal plants. It was invisible schooling, and the result was entire communities connected to their natural environment.'

Jon had unpicked the seamless teaching models of Indigenous cultures and sewn them back together for a modern audience in a set of thirteen practices of 'deep nature connection' — 'deep' because it combines knowledge of place with a felt sense of empathy and belonging. Tracking is one of the practices.

Eva pokes her head into the blackberry bush and looks around for more evidence. I can almost hear the sound of her connectivity-synapses firing, threads of relationship extending out from her to Stampy, the blackberry bushes she's crawled into, the ground she traced with her fingers, the red gum she sniffed the base of. It's ecological literacy not through learning *about* nature, but *from* nature — information absorbed as if by osmosis.

'The way we know it's working is by the sparkle in the eye,' Jon said. 'It's one of the indicators of what I've come to know as the attributes of connection — the qualities that naturally arise when we start to really open up to nature: happiness, vitality, the ability to listen deeply, empathy, helpfulness, true aliveness, gratitude for life, compassion, forgiveness, and the quiet mind.'

It's pure creation mind. Since my busyness meltdown, I've been using my curiosity about bird language as a way to bring me into this space. Tuning one ear for what the birds are saying grounds me in an expanded sense of self that includes but is not limited to whatever task I'm focusing on. I can simultaneously be in the world but not of the world.

Eva brings her head out from the blackberries, dusts off her elbows, and says, 'I think he doesn't want to be found today.'

'Yeah, maybe you're right.' We make a pact to return another day.

'What did you find?' Eva's mum asks, smiling as her daughter spills the stories on their doorstep. She's hardly recognisable as the reserved kid I picked up from the front door an hour or so earlier.

'We almost found Stampy,' she says, her eyes bright and skin flushed rose.

Who really are the teachers of the wild and the untamed? The ones who literally dwell closest to the earth, socks and shoes kicked asunder. Those hiding in branches or bushes, or under tables to eavesdrop on your secrets, knee-deep in muddy puddles without a care for squelching-wet sandals. The ones who take our hands and pull us down to where spider arms and slater legs tickle the ground, who beg us to follow them through moss and mud — *come on, hurry up*, they cry, tugging at us to recall what it's like to hold nothing back, to play and roll and wrestle, to be creaturely and grass-sodden, to hold a paroxysm of sobs and ecstatic joy within the same breath; asking us to consider the possibility of gathering lavender and snails in the same hand. Why stop playing if shivering or hungry when there are lizards to be caught, shelters to be built? That is, of course, and only, if they are *allowed* to be children. If they are let out long enough to tell secrets from the forked branch of the pine tree in grandma's backyard, find fairies in ground hollows, hear bears in the park thicket. Somewhere, anywhere wildish enough to make acquaintance with bird and butterfly, caterpillar and cricket. Somewhere, anywhere out of the adult gaze long enough for the imagination to flesh out character and conversation with elf, grasshopper, and arachnid. Proper nature. It doesn't take much. It's all there waiting in the trembling wings at the back door.

Kids of this ilk are a threatened species, according to Richard Louv, who posited in his bestselling book *The Last Child in the Woods: saving our children from nature-deficit disorder* that the current generation of kids are growing up nature-deficient, and links this to childhood health problems such as obesity, attention disorders, and depression.

Louv is an American, but I see the same concerns in Australia. We're the outdoorsy nation no longer, according to the Bureau of Statistics, which says 98 per cent of Australian children between the ages of five and fourteen watch TV or videos out of school hours as their most common recreational activity. Even the great Aussie backyard is in decline. Until the 1990s, many houses were built with big backyards, but the trend is towards larger houses that cover as much of the block as possible, leaving little room for gardens.

Louv argues that children are being kept 'out of the woods' due to factors such as parental concern about safety, less availability of natural areas, and the lure of electronic devices. This, while also being fed a regular diet of news about the destruction of nature. The average kid in your street, Louv says, may be able to share stories of climate change and extinctions, but have nothing to offer of the last time they made a backyard cubby.

Knowledge-rich but relationship-poor is surely a diabolical combination. And it's not just kids. We're all in this petri dish right now. The good thing is it's never too late to adjust the balance.

The wetland is watching me. Ten pairs of magnified eyes are dotted around the kidney-shaped wetland. Even though I can't see them, I can feel their gaze. The pods of waterbirds splashing around in the shallows seem oblivious. I twiddle a pencil in my fingers and look down at the page in my journal. It's unusually structured, with five numbered rows, each listing time, weather, and activity. The first three rows are already scrawled with bird ID acronyms — 1 x GST (grey shrike thrush) flying east, 3 x DMH (dusky moorhens), 2 x LR (little raven). A criss-cross of lines with arrowheads indicate direction of travel. I look down at my watch. Time for a cooee, but I'm reluctant to break the quietude. The lazy autumn sun is melting me like snow towards the ground, where I would be quickly absorbed by the dark loamy soil of the bank. Still, I create my hands into a conch shell around my mouth, breath deep, and let out a long contact call. A faint echo reverberates back from the tree line on the other side of the wetland. *Period 4*, I mark in my journal. *Calm and sunny*. I rest back against the tree trunk. Looks like it's going to be fairly uneventful on the bird front. I was hoping for a bit of action, but everything down here so far is postcard nature: sweet and benign.

We're a couple of months in now to Rewild Fridays. We had our first go at fire-by-friction last week, with great success, have collected reeds to dry for basket making, made pesto from foraged weeds and wild seeds, started some tracking basics, whittled and carved spoons, and have delved deeper into natural movement. Today, we're exploring bird language: how to interpret the vocalisations and body language of birds so as to learn more about bird behaviour and that of other animals on the land. Baseline, I've told the crew, is the landscape in its normal flow of activity. One thing we'll be looking for is spikes in baseline caused by a disturbance that then ripples out in concentric rings.

I bring my binoculars up to my eyes and scan the wetlands. 2 x EC (Eurasian coots) in the west, Ma (magpie) in the casuarina to the south, flying west. I can see the top of Tas's head to the far south, and nearby Laura in a green beanie half-hidden next to clump of reeds. I pick up the flash of Elena's red scarf from across the water. Only the colour gives her away, sitting still as the water on the pond. She brings the binoculars up to her eyes and struggles to adjust the focal length. I can tell she's frustrated and keeps switching between looking into the binoculars and fiddling with the settings. Finally, she sorts out the focals and rests her gaze on a couple of ducks.

Elena and I had a laugh this morning as she shared the latest from her sit spot. 'A few people have approached me to ask if I'm okay,' she said. 'I now take a keep cup with me. Somehow, it's more culturally appropriate to sit by yourself in Melbourne if you have a coffee in your hand.' The other night she had come across a man squatting and looking intently at something on the ground near her sit spot and decided to ask him if he was okay, to get a sense of his motivation. 'He was probably just like me, curious about something, but my impulse was fear.' Elena is finding it harder to sit still at her desk at work, finding it increasingly meaningless. Her workmates have got used to her desk slowly amassing treasures from her sit spot — 'a whole cup full of feathers, stones, leaves and seedpods, and all sorts'. She told me that it's a reminder to her 'that there's a place I can go where I can just be and that's where there's no expectations on me'. She's also started going to her sit spot after dark. I asked her whether she was nervous, and she replied, 'No, I've developed a relationship with my tree, and I feel safe now.' This woman is determined. I like that about her.

I glance down at my watch. Another five minutes left until the final cooee. It's certainly a different beast doing a sit spot with ten others. The group attention is a presence on the landscape in and of itself. Just as a snake or kangaroo creates a unique energetic track, so too does this combined field of anticipation and curiosity. I wonder how the quantum-physics concept of no objective reality affects the movement of the ecology here, the act of observation itself changing the field being observed. If nothing else, I am paying more attention because I'm going to need to report back at the end of the hour. Accountability is the friend of awareness.

Just as sleep starts to flirt with me, my body suddenly shudders with a small wave of tension. The feeling spreads to a tightness over my chest. It has no obvious accompanying thought, just pure sensation. Could it have something to do with the three pied currawongs skulking in the trees directly across from me? I check my watch and log the experience. *11.48 am: tense.* I extend my awareness out wider, perk up my sun-drenched senses.

My anxiety amps up. A single loud wattlebird croak from the north and then, as if someone has turned the dial down, the entire symphony of birdsong drops to almost silent. I sit up straight and glance around. For a few seconds, there is barely a twitter from the shrubs, just a terrible quiet. And then it erupts. The myna birds explode in a cacophony, joined by an entire circus of screeching rainbow lorikeets. Three ducks launch themselves from the shore with loud squawks. I look up and catch the black silhouette of a small raptor cruising directly overhead before it lands on the branch of a river red gum just across the wetland. The mynas head straight for it, dive-bombing in pairs. The raptor tries to retain some grace of invisibility, raising the odd wing in defence as though the mynas were mere mosquitoes, before alighting to a branch a few hops up.

Action stations! Perfect. Just what I had ordered — a veritable textbook spike in baseline. I check my watch. Time. I cooo-eeee, and a dozen heads start popping up from the bushes, chattering in excitement.

'Whodunnit?' says Michelle, walking towards me.

'The sparrowhawk with the rope in the red gum?' I joke. 'I want to know too!'

Lucy appears from the shrubbery with a smile as wide as a Cheshire cat's. 'Was I just imagining it, or did that just get pretty intense out there before the hullabaloo?'

'It definitely wasn't your imagination,' I say.

'Wow, I'm so excited, this concentric-ring stuff actually works!'

I meet her with an excited smile and am struck again by how a face at her age can be simultaneously that of a child. There's an electricity in her body this morning. She's been becoming more and more childlike over the weeks — delightfully playful, irreverent, and sometimes almost rebellious. The push and pull feels important right now. She's stepped back entirely from working as a physician. It's a decision that's more symbolic than practical, a formal separation from her old identity so she can say yes more fully to the yearnings of the bilum.

We divide into groups and map with colours onto large paper the sightings and movements of the birds in each vigil. There are lots of sightings of 'the tweeting one' or 'little brown birds (LBBs)'. Still, in the awkward attempts to recall the sound and shape of each bird are the important first tendrils of connection. The pooled data from our individual spheres of observation translates into a set of interlocking stories. A territorial currawong in an altercation with a kookaburra emerges from the corroborative reports of Michelle, Tas, and me. Mel and Elena observed the family squabbles of a flock of ducks. Another story emerges when several people report seeing a fox pass at different places, which raises more questions than answers about the relationship between fox and bird.

'And ... the last period, which of course gifted us with the unidentified accipiter ... did anyone pick up a shift in baseline?' I'm curious to see who else tuned in to what I felt before the kerfuffle. I wink at Lucy, knowing she felt it.

'Yes, I felt anxious before the bird came over,' says Laura.

'So did I,' says Marietta, a young bush regenerator who has just started coming.

'Great! This is exactly what I was hoping for — and right before the bell! It's a perfect example of how we can register concentric rings in the landscape. The raptor created a large spike in the baseline rhythm of the wetland — which some of us registered as ripples of tension in the body.'

'And then everything went quiet,' Lucy says, visibly excited. 'I always thought bird alarms would be noisy, but this was the opposite.'

'Yeah, what did you notice?' I ask.

'It was like a tunnel of silence,' Lucy says. 'Followed by the noisy birds.'

'Right, silence first and then a wake of alarm.'

Tas, listening to the commentary, reaches forward to grab one of the bird books and starts flipping through the pages.

'It's like we all have a piece of a giant puzzle,' Michelle says. I noticed she took her sit spot out further today. She told me that she's been going to a sit spot in the grassland near her home and just sitting with the fear — 'letting it wash over me to see what I can learn from it'. Michelle tracks her fear to the societal messages that are given to women about the dangers of being alone anywhere, especially the bush. She attributes her interest in nature connection to her parents, both of whom migrated to Australia for an easier life than the one offered by their peasant farming life in Croatia. Although choosing suburban Melbourne to settle, they carried with them the value of a rural life, and weekends meant family gatherings picking strawberries or meandering through bushland. They also passed down the value of a larger sense of family. Despite the stories of hard winters where food was sometimes scarce, Michelle longs for the community life that her parents grew up within. Coming to these Fridays is a way to reclaim part of her heritage — community and intimacy not just with the land but with a sense of village too.

We circle up on picnic blankets for a late lunch. The buzz of the raptor bubbles forth in chatty excitement. The group feels like it's really found its home. Tas sits next to me, spreading avocado and slices of cheese on corn thins.

'How are you travelling?' I ask, deliberately keeping the question open.

'Yeah, good,' he says. 'That was really amazing, mapping it all out like that. I was imagining if we could make the invisible rings of disturbance from that raptor visible, the land would look like a pond of overlapping ripples. I guess that's what tracking is, hey. Learning how to see both the visible and invisible.'

'Exactly.'

'So this might be a good time to share with you that I've recently realised that the wattlebird that I see around my front yard each morning is the *same* wattlebird.' I stop eating as I register the significance of what Tas has just told me.

'Tell me more,' I say.

'It literally just struck me as the most obvious thing in the world, that the bird I see every morning is probably not a new random bird but the same one,' he says, laughing as he sees my emerging grin. 'And in that moment, I also realised the extent of my disconnection — I might notice a bird but never know it as a particular bird. It's such an anthropomorphic viewpoint that everything is in reference to me or us.'

I take a few moments to think about what he's just said. It's actually quite a profound realisation, a kind of naturalist threshold of Copernican size — the perception of nature shifting from chaotic and random to routine and trackable, from foreign and other to knowable and personable. It opens the first door into a reciprocity: not just knowing wattlebird but being known; not just observer or witness but participant in a relationship.

'So what's changed for you since then?' I ask, already registering from Tas's expression that much has.

'Well … the world just suddenly looked quite different from that perspective. Something about the beauty of knowing that that bird is its own sentient creature, unique from the wattlebird next-door. There's a magnificence to it, the awe-inspiring nature of nature. It's like a world of possibility opened up, like my psyche expanded somehow.'

Tas's face is soft as he describes how he's been experimenting with walking slower and saying hello as he nears the front gate, and noticing that the bird has been shifting its position less far away in response.

'And with this knowledge, I suddenly feel such an affinity for it, like we're both going to work at the same time.'

My heart warms at the story, imagining the morning routine and how the bird might no longer feel as if it's being ignored but instead welcomed in as another home inhabitant with similar routines and needs for food, sleep, shelter, and tribe.

'And my sit spot is different too. I'm approaching it more like my wanders, without any expectations, and it's really beautiful.

This is the genius of it, I'm realising. You have no control of what may happen, and if you can be open to that, you'll never be disappointed.'

I can tell Tas is feeling his way into the next thought, and he speaks it cautiously.

'Something's … happening to me with this nature stuff … and I'm not quite sure what yet. But it's important.'

Taking in Tas's words, I look around the circle at the beings chatting and stretched out like water dragons in the dappled sun. I'm suddenly very aware of a particular quality emanating from everyone, strung through and between us all like spider silk, beautiful and fragile. It's the pure quality of longing.

While Tas starts packing up his lunch, I wander down to the water's edge. I've been downplaying the significance of these Fridays. Partly because I wasn't sure what was possible in a weekly city meet-up, but maybe also because I wasn't willing to acknowledge what I know to be true: that when someone enters into an environment where they're receiving skilful mentoring in the technologies of connection — to nature, themselves, and others — opening themselves up to a relationship with the wilder world, expanding their senses, and repeating this week after week — things *do* start happening. And what are those things? All the unmet longings for connection and belonging starting to wake up.

It's the inbuilt expectation of a toddler to be drawn forth into knowledge and love for the world by their community. It's the hand that unfurls instinctively, expecting the hand of a loving elder to take it and guide us into a life of connection and belonging. It's the assumption that you will be taken by your grandmother into the night and whispered the myths of the stars, that your mother will point out to you the butterfly, and tickle your face with a leaf, and press your hands into mossy rocks, not just on holidays but every day of your childhood. It's the seven-year-old you who just needs to run and explore and take risks and climb trees and bring back cicada shells and feathers and be asked so many questions that you can't help but run straight back outside to take another look at that burrow, that insect. Now you're nine, and your neurobiology tells you that it's time to be given errands of competence, to be asked to go pick herbs and then cook, then stay up with your father and listen as he tells you a story of when he was your age. When you're eleven or twelve, the elders are teaching you how to be in good relationship with the village, how to be a good person, how to have gratitude. Now in late adolescence, it's time for initiation, and you're taken out into the wilds of the mountains to sit on the land by yourself and ask the bigger questions. In this rite of passage, you step out of your youthful social identity and experience yourself as magically and intricately interconnected to the stars and the moon and the trees and the plants and the animals and the ancestors and the unborn generations. You return and take your place as an adult in the village, fulfilling your unique place in a world you know now to be full of mystery and meaning.

For most of us, though, the hand that unfurled in good faith was not met with guidance. Instead, we slept alone, were controlled-cried, learnt how to please, learnt how to sit still in classrooms and recite facts as a marker of our competence. We didn't come to know the plants and animals as kin. We were not mentored to know ways of the earth. We were not encouraged to disappear in the back garden for hours. Instead of a village of eyes witnessing us and asking us questions to weave us deeper into the web of place, at best our village was made of one or two. These two might have been present, but were unlikely to have had the kind of presence we craved or needed, because they too didn't have their longings met. By the time we were teenagers, we had started to close down in resignation and anger. Some of us tried to self-initiate with unhealthy risk-taking and rebellion; others kept small and marched obediently to the orders of the establishment. We don't find much meaning in the empty models of success that mainstream culture offers, so it's no wonder we suppress our longings. What good are they if they are consistently unmet?

Lucy shared a dream with me recently. In it, she discovered a young girl, herself, locked in a box. The girl was squashed and in agony, and Lucy needed to give her mouth-to-mouth resuscitation to save her. Here too, these simple practices — sitting in nature, learning skills, sharing stories — are breathing life back into the longings of the child, are taking them out of the box and saying it's safe to reclaim your true inheritance. The hand is unfurling again, willing to risk the vulnerability of opening.

It's a big responsibility to be one of the hands to take hold of. Because I know from my own experience that when this sleeping giant wakes up, it's with a force and an urgency that attests to the strength of our evolutionary biology, the aeons of our ancestry. This is where I need my own mentors.

The next day, I call Jon from my sit spot.

'You're right,' he says, and I can hear the clink as he stirs ice cubes in his iced tea. 'When you help people reconnect to nature, you activate the nervous system and start to tell it that we're going back to the original longing, and we're going to address all these unfinished bits of business. The longings are not psychological or personal, although we may perceive them as such. You're reactivating that. It's big. The thing is, in modern culture, these longings exist as an unspecific sense of emptiness. People adapt or distract to avoid this feeling. We develop masks, and this inauthentic self becomes who we think we are.'

It's like a kind of ecological Stockholm syndrome, where

everybody unconsciously aligns with the oppressor and is content to stay right where they are and not risk changing anything or making a difference.

'It's socially normalised historical trauma,' says Jon. 'It's essentially uninitiated behaviour, and it's emotionally abusive to the needs of the nervous system.'

'I know, that's what I'm sitting with right now, the enormity of when the dam of developmental impulse breaks.'

'Yes, it's rehabilitation work, but you don't have to *do* anything, Wren,' he says, and I smile at the use of my little-known nickname. He's one of the few people who still use it. 'You just need to let nature take its course. Let me rephrase that: technically, you can't rehabilitate something that hasn't been habilitated in the first place.' Jon goes on to explain how paediatric occupational therapist Kathleen Lockyer coined the phrase. 'So what you're actually doing is helping their nervous systems habilitate.'

I pause to ponder this new word. Habilitate. So close to habitat. Like finding the pathway home again.

'Your year without matches was a journey of habilitation — giving yourself what you'd longed for since you were born.'

At this suggestion, I'm back in a memory of a sitting around the fire one evening with the small tribe I shared the land with, weaving a basket, laughing at tales of misadventure. There was such a feeling of peace. I didn't want for anything.

Jon's right. As I dived into ancestral lifeways over the course of that year, I told my nervous system that those longings that were repressed as a child might actually have a hope. With every bird language experienced, with every bush food I ate, with every track I followed, every sunset I gave thanks for, I reactivated my central nervous system according to some original design. The box of cultural disconnection began to crack. When my connectivity system reached critical mass, I had breakthroughs.

'Your longings were welling up in you with such power and passion that they overwrote everything you thought your goals were, repatterned your motivation system, changed your compass bearing entirely.'

'Yes, that's true, until I hit the wall of grief,' I say grimly.

'Ah yes, true. Well, there's that to look forward to.'

The 'wall of grief' is a term coined by Jon to describe the phenomenon he has witnessed again and again, after mentoring thousands of people. When people start truly awakening empathy with all living things and strengthening the ropes of connection, they oftentimes hit a place of deep grief, realising the generational disconnection that is part of their ancestral legacy.

The end of my naturalist honeymoon period came during my bush year. The first few months were pure joy. A lifelong flirtation with nature was finally given the time and space to turn into a full-blown affair. It was during the depths of my winter hermitage when the grief first bubbled up. While some of it I recognised as old personal stuff surfacing, the majority I couldn't account for. It didn't feel so much like 'mine' but like something arising through me, a product of the process of slowing down. The deeper I reconnected, the more the grief seemed to stir, equal in measure to the continuing joy of the unfolding relationship. It was confusing, like receiving mixed messages from a lover, and yet somehow I knew it was an opportunity for deeper connection.

'As you help these folks awaken, a new field of love and connection is generated. And as that happens, the gift of their own beingness emerges, and with it too the wound of shattered culture. So as they begin to hear the voices of nature again, their hearts will break open and move into this really deep space of sadness and anger at what they didn't receive.'

'Yes, like it did with me. And this time, I'm going to be the one to catch them.' I remember Elena telling me that after the first Rewild Friday she almost didn't return, because the contrast between her experience of connection and authenticity outdoors and her life of office work and isolation was too great — too painful. 'I'm not sure if it's worth it,' she told me.

A noisy flock of lorikeets fly over as if to emphasise the point. Jon continues. 'Nature-connection work is not just warm fuzzies. It's cultural repair. It's relearning a pattern language that goes back to our indigenous roots — tracking and bird language and earth skills and sitting in nature and sharing stories and food and seasonal celebrations. It's village building as best we can — gaffer tape for a shattered culture.'

'Gaffer tape,' I laugh, feeling lighter already, the metaphor endearingly apt. There isn't any magical panacea here. 'So it isn't going to be pretty then.'

'Hell no, it's patchy and sometimes can be a downright shit show, as you well know!' Jon says emphatically. 'The art of building ropes is not the art of instant ropes, it's the art of designing the right routines that get you from strings to ropes. Slow and steady routines of habilitation are what we're talking about. It's the neurobiology of longing.'

'So the pathway to belonging starts with longing,' I say.

Jon laughs. 'Nice one. Yes, exactly. And it can be fraught. People can look to you to give them what they weren't given as a child. But direct them back to their sit spot, back to the storytelling circle, and it will all be okay. Wren,' he says, getting serious, 'western society was born out of moments of historical trauma, often incredibly tragic, destructive, oppressive, manipulative, psychologically damaging, fear-inducing, and disempowering. We are literally trying to help our world heal from thousands of years of disconnection of human beings from nature and culture.'

It reminds me of Richard Louv's findings that many entering adulthood are trying to reconnect themselves to nature, having 'tasted just enough nature to intuitively know what they have missed. This yearning is a source of power,' he says. 'These young people resist the rapid slide from the real to the virtual, from the mountains to the Matrix. They do not intend to be the last children in the woods.'

Jon's face suddenly appears as he switches his video on to give me a smile. 'But really — what else would you rather be doing with your time?' He holds up his glass in a cheers and waves me a goodbye.

Cultural repair. I look out to the swirling, muddy waters of the Yarra. It's true. You can't heal culture without nature. And the first step in healing nature is embedding it back into culture. These Fridays are providing some scaffolding for the crew's longings to take shape, even if the shape's tied together with string and gaffer tape.

I remember back to the last Rewild Friday down at the wetlands. Mel gathered everyone together in a circle at the end of the day and led us all in a round of song. I watched Laura as she lifted her voice and sang it to the sky. It was as animated as I had seen her. Rolling up the bird maps as everyone was packing, Lucy approached me with a cheeky smile on her face and said, 'You know, it's never too late to have a good childhood.'

Chapter 7

You can do it. Come on. You were once a worm. You can be a worm again now. My nose buries itself deeper into the grass as I concentrate on raising only my bum in the air. I'm not meant to have a nose, anyway: it's a protuberance far too evolved for this animal incarnation. Serra, my partner for the exercise, walks her fingers down my back in sync with my slow undulation. In my mind, I am performing a near-perfect worm wave, almost exactly like I saw our instructor Simon demonstrate moments ago. Serra doesn't agree.

'Right there, your back is flat. Separate these vertebrae and get them up. Come on, you can do it!' she says with personal-trainer-like authority, her curly red hair spilling onto my shoulders. I try to imagine the bones rolling in snake-like fluidity, but they won't seem to draw apart, instead remaining glued up together at my lower thoracic, lazily accustomed to human bipedalism.

I groan under the strain of holding myself up for so long. I'm determined, though, not accepting yet that I haven't quite got what it takes to be an earthworm. I did, once upon a time, Simon reminds us.

'Each of us carries the entire history of our ancestry inside us, from our primate heritage, stretching back through our reptilian and amphibian past, to when our family were fish and worms, and then all the way back to single cells floating in the ocean. So if I want to gain full awareness of the positions and potential movements of each of the twenty-five or so bones of my spine, I need to focus my attention as intensely as possible on each vertebral segment individually.'

I sneak a peek at Simon. He's calmly chatting away to us while hanging off a branch with one limb. He looks like he belongs more in a tree than on the ground. A friend from up north, Simon is camped down in my backyard while he's in Melbourne to run his Ancestral Movement workshop, designed to explore the patterns of movement and awareness encoded in our bodies — starting, it seems, at the worm.

'As we explore, we find that the body is full of layer upon layer of extraordinary, ancient, ancestral power. Four billion years of adaptation and embodied knowledge. It's all a bit epic,' Simon says by way of encouragement.

'What's epic is that I get to roll in the grass with babes,' Serra says in my ear, clearly distracted by some of the other fine-looking prostrate primates nearby. I stifle a laugh with my hand before remembering that worms don't have them and glue it back to my side.

My body aches with the effort of worm, but I'm not giving up. I lift my mouth to the sun, feeling relief to be at the penultimate shape of this single cycle of forward motion. There is nothing but my arched body soaking in the warmth, and for a moment I sense what it might feel like to be a single-tubed organism, muscles contracting in waves of instinctive propulsion motion, my universe made up of food and sex and not much else. I collapse on the grass, my spine tingling.

'Again!' says Simon. 'This is where it starts, with the spine, capable of so much varied movement. Imagine yourself a fully embodied human, leaping and jumping, vital and vibrant for the rest of your life.'

This is enough to motivate me into another worm wave, while Serra gives me a commentary on the size of the biceps on the worm opposite.

With some relief, we soon skip some millennia and evolve into other variations of spine movement: lying down, sitting, kneeling, squatting, standing, rolling, crawling, walking, running, jumping, climbing, balancing, lifting, throwing, wrestling — just to name a few. We shift between 'earthquake neck drill' to 'crocodile drill' to 'balance battle', all of which have the common goal of contorting my body into unfamiliar shapes.

The park becomes a training zone. Every safety barrier, fence, or stairway designed to direct human movement in a specific way is reimagined as an opportunity to facilitate nonlinear animal movement. We move up the steep stone stairway like lizards, keeping our long bodies taut and parallel to the ground, our spines supple and swaying as our long claws move in syncopated rhythm. A treated pine barrier is the place to crawl, slide, jump, and balance one-footed. Trees are an invitation to morph into possums, koalas, chimpanzees. We swing like a tribe of sloths from branch to branch with non-verbal hoots and grunts of joy.

In the grass amphitheatre, we roll and tumble in slow-motion capoeira cartwheels, getting as close as we can without touching. It takes control. Eventually, we make contact and morph into two- and three-headed creatures, moving as one. Floating over someone's back, I look up to see an ibis fly overhead, and I become even lighter, the mirror neurons in my brain yearning to enact its seemingly effortless movement — I feel myself pulled up towards it as if a kite string is attached between us.

A small crowd gathers, watching the primates in a zoo. Even the magpies inch in curiously.

Simon stands stork-like on one leg and seizes the opportunity to make a cultural comment. 'Much of the writhing and undulating movements of the body's central axis that we are exploring, such as the worm, are taboo in our society, only acceptable during sex, away from the public eye. It's considered strange to do much else apart from stand, walk, run, or sit.' Some of the onlookers shift a bit uncomfortably, as if embarrassed by their upright gait.

It's true, though: I often attract strange looks while climbing a tree, or even something as simple as swinging my arms in a circle.

'Our animal bodies are actually capable of an infinite number of movement possibilities and expressions.'

Simon leads us down to the creek, where a thin metal railing borders a steep set of bluestone stairs. He alights the railing with a single jump and proceeds to pad his way up the beam with feline grace. He signals: my turn. I jump up on the cold beam and cling to it with hands and feet, wobbling dangerously to my left where a two-metre drop leads to a muddy bog.

'It's easier when you start moving,' he says.

I will my hands and feet to become agile paws. My front paws grip the rail white-knuckled. How does a cat do this? What follows what? I try to remember the gait that I've seen so many times before, but it's not coming to me. The urge to shift my weight backwards into two-leggedness is strong. This is not a yoga class. There's no mat and no formula. Small muscles fire up in my calves and core. I pick up my front right limb and launch it forward a single step and am amazed when the opposite back limb follows. My toes push me forward again, and the left front paw moves and is met by a rear. That's it! Cat, dog, fox. Somehow, I instinctively know this diagonal gait. I can feel Simon's eyes on me and will myself to keep moving rather than look up. My next few steps are fluid and confident.

All of a sudden, an image of a puma comes into my mind. In that split second, I am simultaneously seeing through its eyes, the detail of the ground below and the trees above popping out in high definition. The earthy scent of mud is strong in my nostrils, and the wind through the leaves is stereophonic. I pause, feeling lithe and limber, supple and defined. Ready to pounce. In this launching position, up here on the rail, I suddenly know myself to be practising something very old and very valuable. I take a few fast and bold leaps along the railing and jump off, to Simon's cheers.

Sitting down at the top of the stairs, I watch the others attempt a similar shapeshift, with varying success. There's a mix of frustration, elation, and camaraderie. A feeling of empathy for the big cat remains in my body. I think back to the story Simon shared yesterday as we drank tea on the deck, about how he started exploring animal mimicry.

'I spent days one summer quietly following a monitor lizard around in the forest, feeling clearly for the first time that the undulating spine and alternating step in its crawling pattern were the same as those in my walking and swimming. Not just a head, with its eyes and ears and mouth and nose, a neck and a spine, but hips and shoulders, elbows and knees, wrists and ankles, fingers and toes. Most of the fundamental patterns of my own body structure were present in that lizard, which, to me, meant that I was in many fundamental ways still a lizard.'

'How did this affect you?' I asked, noticing the broad spread of Simon's toes as he placed them on the deck beam.

'There was an intimacy, an understanding. Like some gap closed.'

Mimicking the movement of animals is a practice found in cultures around the globe. Many of the martial arts from Asia are based on imitations of crane, tiger, or turtle. Indigenous Australian cultures regularly adopted animal forms for ceremony with the help of body paint or masks. Cave paintings and stories in Europe indicate that these ancestors did the same. I think back to the local Indigenous dances I've seen — the subtle movements enough to powerfully evoke cockatoo, heron, emu, eagle, wallaby. This shapeshifting is a way of relating the patterns and connections of the living world, embedding the knowledge of animal behaviour and characteristics in the body until it literally becomes embodied. From this place comes an empathy and understanding, a deeper connection. For a hunter, this is vital, not only to understand the physical tracks, but as a gateway to the mind of the animal itself, and therefore to survival.

At the end of the day, we finish where we started, horizonal to the ground, this time on our backs, shoulders touching, heads meeting in the middle. I imagine us like a multi-tipped starfish, belly up to the sky. Our collective scent is pungent and indiscriminate. We're covered in grass scratches. Muscles I didn't even know existed are aching. Gravity is a large bear, its heavy, warm paws pressing me like a leaf into the earth. Simon invites us to close our eyes and focus in on all the sensations within our skin. Although I'm so tired I couldn't bat away a mosquito, I'm also tingling inside and out, as if little lights are switching on all through my body. I feel more alive than I can remember.

'Our perception of time and space shift. We feel the fact that we are giant organisms of mind-boggling complexity, made of water, rock, and air. And more and more, we sense and feel the immensity of past aeons right now, in the present moment.'

I expand my senses, and the tingling amplifies until my entire body is buzzing so fiercely that I'm surprised the person next to me doesn't feel it. The boundaries of my skin become indistinct, the buzz expanding out beyond me to mingle with the tingles of the group, the trees, the grass, the ducks, the streets, the neighbourhood. All is vibrating energy. In an uncharacteristic cosmic invitation, Simon encourages us to beam out love through the aliveness of our bodies. It's an easy leap, to imagine this expanding, tingling body as enlivened by the quality of love. I send it out as ribbons of loving energy into the realms above, below, and in between. It comes back to me like an echo of delight, and I soak it into my core, from where it spreads, butter warm.

'Our deepening sense of ourselves, our minds and our bodies, grant us a deepening sense of the living world and our continuity with it, and eventually, at a certain point, we come back to a very simple and natural form of worship of life itself.'

We rest on the earth in the sweet bliss of cellular aliveness. Nothing more. Nothing less.

The next morning, I take my stiff muscles for a stroll down to Dights Falls, where Merri Creek meets the Yarra. Despite the highway noise and collection of plastic water bottles, I can easily imagine this as the gathering place it once was. Ducks swim like quiet rafts in the mist. A carp breaks the still surface and arches its entire body out of the water. Wow — what strength it must take to crest like that. My spine spontaneously shifts into a small mirror likeness of the spinal arc. Typically, I might resent the presence of the introduced carp, due to ecological considerations, but watching it today I am in awe of its power. It comes down with a splash and disappears, but I follow it with my imagination, its spine rippling with water-like fluidity as it makes its way upstream. I yearn to join it in its joyful morning routine, but instead I squat at the water's edge and tune in to the tingling sensation still palpable in my body from the weekend. Waves of subtle pleasure still pulse through me. Even the sore muscles are satisfying. This body just loves to bend and lift and bear weight and locomote. It wants to be fully used. Simon's right: the pleasure of a body extended is a doorway into the pure pleasure of being alive, a direct source of connection to the living, breathing, pulsing world itself.

When I lived out bush, this was a given. Exercise was an irrelevant verb, synonymous with life itself. Months cutting, hauling, digging, bundling, thatching, and weaving created my home. My fire warmed me three times — first in the collecting and chopping of the wood, then in the rubbing sticks together to produce a coal, and finally in the flames. My arms carried every litre of water I needed. I made clothes with the elbow grease required to soften and tan animal hides, I sawed and whittled my own furniture, I dug my own latrine. My feet took me where I needed to go every day, and returned with pockets of berries, roots, shoots, tinder, and plant fibres. I could scale trees, run kilometres, walk for days, balance like a cat, and sleep as well as one too. Sitting was reserved for rest, observation, and contemplation.

Here in the city, the equation is flipped. Fire is the flick of a switch, water the turn of a tap, transport the turn of a key or swipe of a card. One sits for the bulk of the day. Exercise is an extracurricular activity. With some frustration, I've succumbed to buying a gym membership, purely so I can lift heavy things and swim long distances. At one strength class I attended, the instructor said, 'These are the muscles you'd use if you were using a bow and arrow.' Well, why wasn't I? My bow sat unused in my room.

I had to laugh when we installed an electric heater that mimics a log fire in our lounge room. We call it 'Howler'. With the flick of a switch, the plastic logs pulse a surprisingly satisfying ambient warmth into the room, especially when Min adds a crackling fire soundtrack from YouTube. I curl up in front of its breath of hot air from the vent and reflect on all the things that are being lifted or transported without me in direct service to my survival needs: trucks burning fuel to cart food grown miles away, power stations fed black rocks hauled from the earth, water dammed and pushed through pipes to reach my tap, machines harvesting cotton to make my clothes, trees cut and hauled to make the paper that is the napkin under my cutlery at the cafe. Life requires heavy lifting. It's good to know the real weight of these things. And in the city, we hardly lift a finger, having outsourced the effort to others, including the future generations we are indebted to as we burn the last of the ancient sunlight in oil.

The fossil record shows it's taken around six million years for *Homo sapiens* to evolve into our present form. With the exception of the past two hundred years of the industrial revolution, our ancestors were active daily simply in order to exist: first as hunter-gatherers and then working in agriculture or in jobs to support it. Enter the hyper-technological office era, and *Homo erectus* has turned into *Homo sedentarius*: persons that sit.

I go to a co-working office a couple of days a week. On a productive day, I might stay on my laptop for four or five hours. Still, I shake and stretch and moan. Others seem to sit unmoving for hours on end staring into the screen. How do they do it? How do their bodies not scream and tantrum in protest at this incarceration? These are creative minds and wildly sensitive bodies hunched over in expressionless immobility. Is this really the best we can imagine for ourselves?

Sometimes, the only medicine is to get out in it. Really in it. Sometimes, as Martin Shaw says, in this era when we have central heating and 'twenty thousand different types of bread within a stone's throw', we have to choose to trade shelter for comfort.

'There's no way I could sleep in that!' exclaims Michelle as she watches Laura shuffle feet first into a corridor of sticks just large enough to fit her small frame inside. Laura wriggles down so that her head is just inside the two interlocked Y-sticks at the entrance. The A-frame shape is shoulder width at ground level, tapering down to a narrow funnel at the feet.

'My toes are tickling sticks,' Laura says, pushing out a few of the sticks with her feet. Her brown curls and green top are camouflaged against the floor of leaf litter.

'Can you roll over?' Elena asks.

Laura gingerly shifts to her left side, her right shoulder barely shy of the top of the frame. 'Just,' she murmurs.

'Its affectionate name is "the claustro",' I laugh. 'But it's a survival shelter, remember, not something you'd live in.'

'Time for leaves?' Lucy asks, with a tarp already in her hand.

'Yep. Leaves, leaves, and then more leaves,' I say.

Laura shimmies out of the shelter, and the group gets busy raking and scraping debris from the ground onto small tarps and tipping them over the shelter's frame. The design is the most basic of survival shelters — a stick frame just large enough to scaffold the body, which is then insulated with piles of forest debris. The leaves act like a sleeping bag, trapping warm air in the gaps. We've chosen an out-of-the-way place in a park bordered by spiky bushes and under the cover of a small eucalypt plantation, which affords enough leaf litter for the roof. Two hours later, the shelter has amassed enough leaves to keep someone alive, if not warm. The group stand back to admire their handiwork.

'It looks a bit like a dinosaur burrow,' says Lucy, wiping the back of her hand across her face.

'It's actually really beautiful,' says Michelle. 'Like earth art.'

'Anyone tempted to stay the night?' I ask jokingly.

'Not I, said the fly,' says Michelle. 'Can you imagine the bugs?'

Laura stands back looking at it thoughtfully but says nothing.

'Hey, look, a nest!' says Elena.

'So it is!' says Lucy, and we follow her to the edge of the prickly moses, where we can see a small clay and mud nest, perfectly protected in amongst the spiky thicket of leaves. A scrubwren flies in and flits around, obviously distressed by our attention.

It's a small group today, and the half-dozen of us wander down to the river and spread out rugs out on the bank for lunch. The chatter is buoyant, faces streaked with dirt, and hair strung with twigs. As I bite into my sandwich, I notice the dirt etched in lines on my palms.

'Why is this stuff so satisfying?' asks Lucy. 'I mean, all we really did was make a huge pile of sticks and leaves. I didn't need six years of university to equip me for the task, but I'm loving it.'

'Same with fire-making for me last week,' says Elena, who sits a little further back from the group so as not to be tempted by food. She's begun fasting one day a fortnight in order to prepare herself for a four-day fast as part of a Vision Quest at the end of the year. 'I've been spinning the sticks between my palms at night. It's keeping something alight.'

Last Friday, we made fire with a hand drill, employing the Indigenous method that uses one stick — in this case, a grass-tree flower stalk — spun with enough friction onto a base board that's notched to size. To begin with, I had everyone whittle their own kit before splitting into small groups to make fire. I crouched next to Elena's group and encouraged them to get a feel for the stalk moving between their palms. I remembered my own first time: it looks so easy until you try to coordinate the speed and pressure necessary for smoke to appear. I held back the urge to help as the group rode waves of exhaustion and hope, as smoke appeared and disappeared, whispering its promise, seemingly in equal measure unattainable and potently close. Their first attempt failed, drilling through the board and into the ground. Elena wasn't going to give up, though, and roused her two companions to try again. Despite the crisp autumn day, sweat dripped from their foreheads by the time a small red glowing ember rolled out of the notch like a tiny snake. Elena could hardly believe it, and I guided her to cup the coal in the tinder bundle and gently blow. When it erupted into flames, her smile was a burst of golden fire from her soul itself.

'Why do you think it is?' Mel asks. 'Why do we keep showing up week after week?'

The group munches on the sweet potato chips I offer around as they contemplate the question.

'There's something for me about using my body for a purpose that feels meaningful,' says Elena. 'Sitting all day drains me.'

'It's just so real. So much is conceptual these days, and you can't argue with fire or food or shelter. It's authentic,' says Lucy.

'I like being uncomfortable sometimes too,' says Laura. 'Makes me feel alive.'

'Me too,' says Elena. 'That's why I've been doing the cold showers and ice baths.' I've noticed a shift in Elena since she started the cold exposure and the fasting. She's got more spine, more substance.

'I've noticed that my sensitivity to clients has changed,' says Tas of his osteopathic practice. 'Since sensing more in my body, I'm sensing more about others.'

'But there's also fear of pleasure too,' says Michelle. 'Like some guilt kicks in that tells me I should be more productive than just sitting here soaking in the sun and observing nature.'

Post-lunch, the conversation trickles down into a contemplation of the sound of the water bubbling past. We're all just standing to head back when Laura says with quiet assurance, 'I'm going to sleep in the shelter tonight.' The resolve lands with the splash of a rock in the languid group pond. Questions rain down. *What, really? But you don't have warm things. What if you get too cold? It's forecast to rain tonight. You've only got your bike, and it's an hour's ride home. What about food? What about strangers?* Laura sits back, listening, and fends them off with a single statement: 'The soul grows strong when the body does too.'

Once the group feels Laura's commitment, the energy shifts into a motivated responsibility. There's talk of adding more leaves and constructing a door plug that Laura can pull in behind her to keep the cold wind out. I recommend smudging the inside of the shelter with smouldering green gum leaves to deter bugs. Survival practice has suddenly turned real: it has to keep one of our tribe warm. We head up the hill with purpose in our stride.

When we farewell Laura, a southerly breeze has blown in, and I wonder how she'll go without so much as a blanket. We're leaving her with a supply of snacks, having topped up her water bottle and showered her with blessings. 'I trust the earth,' she says as we cluck around her. 'And I trust my body.'

As I ride my bike into a headwind, I think about how it's really one and the same: earth being body, and body being earth. Trust in one is trust in the other.

Laura's comment reminds me of my friend Meryl, who moved to Melbourne a couple of years ago in order to be closer to her daughter, after thirty-five years living in a rural village on the edge of a river. I was shocked when she told me she hadn't lain on the ground since arriving. 'I don't trust it,' she said. How many other people don't trust they ground they live on in the city, so much that their bodies never fully make contact? We were both at a small gathering on the banks of the Merri, and at lunch I watched Meryl walk down to the creek so slowly as to be almost still. In her hand, she held a keep cup. I watched her enter the water, red toenails bright against the grey of the rocks. She filled the mug with water and, bringing it back onto the bank, slowly tipped it over her head and onto her crown. The water fell over her face and down her neck. She let it run until it was just drips, like the last rain after a storm, then followed the water down to the ground, her limbs making contact little by little, until she was completely flat on the earth. With a deep sigh, she finally released control. I could almost feel the relief as she surrendered, as if greeting an estranged friend. It takes ground to find ground, water to find flow.

The intelligence of the body in relationship to the land is something I keep in mind when we sit around the fire on Fridays whittling traps, tanning fish skins, constructing water filters, twisting reeds into string, flint-knapping rocks into sharp points. Dog walkers wander past and gaze curiously. Why are we choosing to do this? If we're not prepping for an apocalypse, then what are these activities offering us and the world?

Yes, there's the raw authenticity of handling fibres, clay, wood, stone, and bone. There's the joy of simply being active and outside with others. But I suspect the degree of gratification has something to do with the employment of the body — its dexterity, strength, and sensitivity — to tasks directly related to the continuation of life. A walk is good anytime but offers another layer of satisfaction when it's taking me to and from some place I need to go. A tomato grown and eaten from my garden offers a satiation entirely different than one I've had no hand in. The act of carving a fire kit with the real promise of using it to warm myself contains gravitas and meaning far beyond stick whittling. Because it's part of our Original Instructions. Buried in our genes are the ways and means of self-reliance and sustenance. It's the pleasure of remembering something we were all once literate in. We pick up the weaving fibres and the fire stalk, and some part of us recollects this twist of the fingers and press of wood between palms. Our muscles and bones have grown up on these fundaments. Shelter, water, fire, and food. As old as dirt. As real as mud. The body keeps the score. It tells me this is true. This is good. Life-affirming. Enlivening.

But what about the endless emails and bills? How can I integrate body, work, and survival into urban life? How might I find ways to engage the full capabilities of the body as a dynamic perceptive intelligence rather than a vehicle for the brain to be moved about on? If socialising, exercise, work, solitude, and survival (primarily food) are all separate activities then it's a game of moving around competing pieces in a puzzle that never has enough space. But what if they're not mutually exclusive but mutually supportive? At one end of the integration spectrum might be our foraging forebears, for whom these categories naturally blended. At the other might be many of us outsourcing 100 per cent of the labour that supports our life to external and anonymous sources. I want to find more possibilities along that spectrum where the boundaries begin to blur. It's a life-stacking exercise — to see how much I can overlap in the middle of the Venn diagram, and then to live more from that sweet spot.

My friends Meg and Patrick live a life for which they've coined the term 'neo-peasantry', which aims to bolster this sweet spot. It's about cultivating informal subsistence economies of homeplace and community and fostering accountability to their own food and energy resources. I visited them and their son Woody recently to learn more. As always, I gravitated towards the large kitchen in the table for Meg's preserves and ferments. There were some recognisable shapes — garlic cloves in a dark syrup, apple peels floating in a jar labelled 'wild apple cider', baby cucumbers swimming alongside yellow mustard seeds. There were burdock, parsnip, and dandelion root beers, stinging-nettle cider, peach-pip vinegar, krauts and slaws, pickles and lacto-ferments. Small bubbles rose to the surface of seemingly stagnant ponds, their work taking place in invisible alchemy over weeks or months. Outside, the backyard was a neat assortment of productive gardens, tiny houses, a home-built sauna, chickens, bees, humanure bays, covered and split wood, and various other homesteading projects.

They've transitioned from 100 per cent reliance upon the monetary economy to just 30 per cent in the space of a decade, gifting and trading with around eighty other households and individuals and gradually reclaiming the skills of their land-encultured ancestors.

The 'neo' in neo-peasantry refers to the *choice* to be peasant-like in the way they goatherd on public land to mitigate weeds, grow and cellar food, collect firewood from around town, make useful the discards from affluence, bike rather than drive, forage for weeds, herbs, and mushrooms, celebrate the seasons with songs, poems, and community dinners, and generally, as Patrick says, 'live a dirt-rich life'.

Their efficiency pertains to the deeper project in which their lifestyle is embedded, a cultivation of what Patrick names as an indigenous relationship with the world. This tending to the 'spiritual ecology' of life is like the mystic arm of permaculture — little spoken, less practised. It's not a by-product of growing food, but an intentional entering into the mind of the living earth. It's part of their 'economies of belonging' concept that signifies the holism of their worldview.

'If we reconnect to our indigeneity along ancestral lines, then it's not such a giant leap to meet First Peoples of the land in which we live. If we're living in that place of intimacy with our local earth, and inside First Peoples' law, then we're not going to relate from a place that romanticises progress culture. We're both meeting in our own indigeneity, and thus honouring of circular — gift making, giving, receiving, and returning — economies. That's a place to have a conversation from,' Patrick said.

It reminded me of what Tas had told me about his solo wanders stirring him in ways he hadn't anticipated. 'I've always been someone who goes by the rules,' he'd said. 'I would never go off the track, figuratively and literally. But wandering is shifting all that up. The only way I can describe it is that I've become aware that there is an external force present when I'm in that attentive wandering mind. It's related to the land, and there's a sense of being guided. When I'm off-trail, I hand myself over to something that's very powerful and feels very old.'

Perhaps it's as Toko-pa Turner says in *Belonging*: 'Most of us think of belonging as a place outside of ourselves, that if we keep searching for, that maybe one day we'll find it. But what if belonging isn't a place at all, but a set of skills, or competencies, that we in modern times have lost or forgotten.'

As dusk falls and I'm warm inside with my hands wrapped around a mug of tea, I'm thinking about Laura. The building of your own shelter by hand is a skill of belonging, the most basic way of belonging. There's something profound about truly feeling, perhaps for the first time, the weight of your body as it lies on the larger body of earth without any padding between. It's skin-to-skin contact: intimate, bonding. I've done it enough myself to know the power in it.

Sean has built himself a writer's studio out the back of his rental house. He tells me it has a large-enough window and just the right thinness of walls to remain somewhat permeable to birdsong and the weather — though this means it's also draughty. I've been sleeping in a swag by the river some nights. Small gestures can make a big difference. Tonight, though, Laura is truly trading shelter for comfort, sleeping in a stick shelter in a city park, knowing there's some gold in the act of even a small ordeal. As the first stars appear, I send her my best wishes.

A few days later, Laura emails the story.

For a long time I avoided getting into the shelter by wandering around watching the kangaroos, the sunset, the moon, the first stars, the wind dancing. When I finally did get inside I entered into another state — 100% of my senses on. I was literally touching the dirt with my face. I was part of the night atmosphere of the forest, aware of the nearby birds' nest. I didn't feel scared, I actually felt very alive and awake. I was cold and I was shivering and yawning. It was nearly impossible to sleep. I was claustrophobic and restless, nervous and excited.

When the rain came, thanks to all of you the shelter was totally waterproof. In the very early morning, maybe 3 am, the magpie called. I had been inside for about 8 hours, pretty cold, without sleep, just resting. A couple of hours later I emerged dry, while everything around me was wet.

I rode back in the dark following the river and stopped to watch the sunrise at Banyule swamp. My awareness was fully open and full of curiosity. I didn't want to leave the wild bundle of nature. You were in my thoughts, your eyes in the circle before my departure, the warmth of this tribe kept me warm inside.

I get to work on my life-stacking. I up my garden capacity, lifting and digging to prepare beds and plant seeds. Chopping wood takes priority over swinging kettle bells. I find a new half-hour jogging loop that takes me around by the bat colony and the mushroom-foraging field. I get up and outside at sunrise. Walk to the shops and only bring back what I can carry on my back.

I find ways to be more mobile as I work. A standing desk appears at the co-working office and by popular demand soon grows to a fleet of half a dozen, one which I claim. 'It's a small rebellion,' says fellow co-worker Bodhi, who often joins me in pacing around the kitchen and bemoaning sedentarism. I put on my headphones and wiggle around as I answer emails, put my body in unusual shapes throughout the day, and stretch when the kettle is boiling. I start adding my own small rebellions — work meetings conducted as river walks, phone calls similarly scheduled as portable.

Instead of catching up with friends at a cafe, I send out social invitations to join me in the garden or out foraging. I jump stairs and swing off branches as I walk, trying to find new ways to move, climb, and balance, even washing the dishes with one foot off the ground. I beg, borrow, or steal the opportunity to move my body whenever I can get it.

I'm also dipping my toes into some 'biohacks' — a rewilding buzzword for DIY physiology experiments that attempt to up the body's resilience and wellbeing by doing things our ancestors might have been familiar with. Wim Hof has popularised the power of exposure to cold. Like Elena, I've been leaving the cold tap on in the shower for a minute after I turn off the hot, and sometimes diving into the river. I've been told it increases metabolism, decreases cortisol, and increases serotonin. Certainly feels good afterwards.

Diet is the most obviously hackable element. A friend has gone on a high meat and greens diet and only eats once a day. Despite my scepticism, her allergies have all cleared up. 'Three meals a day is a postwar invention,' she tells me. Min, Megan, and I decide to biohack our diet, we embark on a five-day fast supplemented with some veggie juice and chicken broth. Instead of joining, Tully says he'll eat enough for the three of us.

After two days of being fairly hungry and grumpy, I wake on the third day full of energy; we take ourselves to the bouldering gym and climb the holds like skinks. On the final day of the fast, I wake with the birds and trip-hop down to my sit spot. My awareness has gone supersonic, as if my receptor cells are on steroids. I'm sharp, clear, and strong. We are all so overfed! How much lethargy and cloudiness is actually due to the amount and diversity of food we put into our systems? Three solid meals a day is a great consumer strategy but not so good for my digestion. I'm going to start experimenting with not eating until midmorning and then again in the late afternoon. It's all I need.

Another couple of biohacker friends have been climbing to a high point nearby every morning and stripping off to tan-through swimwear to expose their bodies to the light of sunrise, a practice meant to align circadian rhythms and supercharge mitochondria. I join them one morning, rolling up my sleeves and jeans as they get almost naked in a downward-facing dog. My 'is it working?' question is met with an enthusiastic yes, Kara informing me that she now gets tired at 11.00 pm rather than 1.00 am or 2.00 am, and she has more energy.

There's plenty of fads around, but things get simpler when I ask myself the question of what makes me feel good. I need to be content right now with the crib notes from our evolutionary diaries rather than the entire textbook. Make contact with earth and sky as often as you can, be outside when you can, follow feasting by some occasional famines, eat nutrient-dense wholefoods, move your body, give it some challenges, and most of all, listen to what it needs, even, or especially, if that's sleeping in an earth shelter.

Chapter 8

The tube train jerks suddenly to the right, and my right thigh rips up like a bandaid from where it was stuck to the seat. I forgot how unpleasantly stifling the London Underground is in summer. It's like an old carnival ghost train. My jetlagged state only amplifies the sensation of being suffocated.

The lack of phone reception doesn't create the congeniality I would have expected in such a shared ordeal, and my warm glances are met with groggy, half-closed eyes. This is obviously where people catch up on their sleep in this city. At last, the train pulls up in Leyton, the East London suburb where my friend and old housemate Arian is now living.

I spot Arian from the top of the escalator — he stands a good head above the crowd — and I walk towards him slowly, savouring the anticipation of greeting this treasured old friend. He's looking even more Gandalf like since the last time I saw him, his beard wispy and silver, the lines around his smile a little more etched, his sky-blue eyes sparkling with childlike mischief. I watch as people steal furtive glances on their way past. Finally, he sees me, breaks into a huge grin, and begins waving one long arm. I drop my pack and run into his arms, which wrap around me like two pythons, pulling me in tighter.

'Oh, Wren — so good to see you!' It's been some years since we've seen each other, and a good five more since we shared a house in the forest. Tears come to my eyes at the affectionate sound of my old name and the memory of that very tender period of my life when our lives overlapped, me having just emerged from my bush retreat, my wings still tremulous and wet. He knows me in a way no one else does.

Arian slings my heavy pack over one shoulder with ease, takes my hand in his large clasp, and leads me out of the station and around the corner to a neat row of slightly dilapidated old townhouses, one of which is his.

'You must be exhausted,' he says. 'Tea or shower first?'

'Tea, definitely,' I say despite feeling like I have three continents' worth of grime on my skin. London is my landing pad for a week before I head to Spain to teach a two-week Wild Camp at the invitation of friends at the Eco-Dharma Centre, in Catalonia. Visiting Arian was the sweetener that clinched the deal.

The tiny kitchen's double-glazed windows look out onto an equally small backyard, occupied mostly by a treated pine deck that backs onto four postage-stamp-sized yards. A single small tree sits at the back of Arian's yard, casting scant shade. Arian is barely a foot short of the kitchen ceiling, making him appear somewhat of a giant.

'Oh, your teapot!' I exclaim as Arian heaps tea leaves into the china pot that used to live in our kitchen. I would often sit cross-legged on the square red-velvet cushion that took up a nook on the large wooden kitchen bench and watch as he prepared tea. Boil water in the pot, add two and a half measured scoops of Earl Grey, chat while it steeps, turn the pot clockwise twice before pouring. I'm happy to see the ritual has endured the continental move. After nearly twenty-four hours in the sterile and grey world of airline travel, the familiarity is comforting.

The house that we lived in together was very different to this one. The kitchen alone was half the size of the ground floor of this townhouse. I was renting the self-contained west wing of the house. My bedroom had glass walls on two sides opening to rainforest. Every morning, I would wake to the tap tap and meow of a catbird on the window trying to beat up its own reflection. We'd take our tea outside to the sunny couch and listen to the bird chorus between spacious sentences. Most days, you could hear the creek running over rocks, and we'd swim in and drink from that waterway. I couldn't believe I'd found such a place. All I knew when I finished 'the year' was that I couldn't go home and needed to try and put words to the experience. When mutual friends suggested Arian's place, it seemed perfect. It was, but not in the way I was expecting.

Soon after moving in, I set up my desk at the window and diligently sat down every morning with the innocent assumption that my story would pour forth. I tapped the pen. I twiddled with the keyboard. I mind-mapped on reams of butcher's paper. And grew more confused. One part of me knew that something extraordinary had just occurred, and another part was trying desperately to disregard any significance and abandon the crazy compulsion I had to write about it. On top of that, I was totally disorientated, grieving the loss of my shelter, my sit spot, my tribe, and that whole life of wild adventure and timeless wandering. I felt suffocated by four walls and deadened by my attempts to sit at a computer. I soon found out that Arian had undergone a similar immersion, spending some years in the forest alone, was just finishing a PhD, and had a penchant for human psychology and narrative. My own small kitchenette went unused once I found the red-velvet perch and the companionship. Moving in with Arian felt like a key turning in a lock.

Down in the paddock near the creek, Arian set up his usual home-cation summer camp: a self-made big-top-style canvas tent held up with hewn sapling poles and pegs. A swag sat under the canvas and beside the open fireplace, which was tended with homeblacksmithed tools. A set of matching wooden crates stored basic dry foodstuffs. It was bespoke and beautiful.

Arian invited me to share my experiences with him. So over twelve evening fires in the summer of my return, Arian received the full twelve months of my story. He would poke the fire as I read from my journal and reminisced, listening more attentively than perhaps I'd ever been listened to before, not saying much more than *Is that right?* or a long, drawn-out *reeeallllllllyyyyy*, a bit like the crow cawing in the paddock, but with lots more beard stroking. At times, he would interject with questions. When I was done, he would ask a few more questions, stroke his beard some more, pause, and then out of nowhere pick up the seemingly disparate threads and weave them together into a narrative tapestry that was far richer and more beautiful than I had imagined. He took what to me was a purely personal story and cast it in mythological terms. Through his eyes, I heard the story of a young woman who chose to leave her village and enter into the wilderness of her inner and outer landscape. Along the way, she found allies in the birds and trees, became skilled in self-reliance, confronted fears and her own demons, lifted the lid on unfinished business from the past, and finally laid claim to herself and her gifts as a woman. It was the story of my own self-designed initiation.

Through my speaking and being witnessed, the experience settled in a new way within me, like leaves finding their place of rest after a storm. I began to understand why stories must be told. I began to write. I was devastated when Arian told me he was moving to London.

'I've lived in the forest for thirty years. I'm bored,' he said.

I protested but understood too. It was selfish to keep him all to myself. Still — London?

It's a big pond for a tree frog like Arian to swim in. I'm curious to see how he's been faring. I follow him and the teapot to the deck. Loud trucks rumble past, but he seems unbothered. A patch of dirt in one corner of the yard has the remains of some charcoal. Arian follows my gaze.

'I tried having a fire, but it was too awful — neighbours on all sides watching me.'

He splashes the tea with soy milk and stirs in a generous teaspoon of honey. It's amazing how tea can taste so specific to one maker, I think as I take a long slurp.

'Look at us — two city slickers now,' I say as a loud plane flies overhead.

'I know, who'd have thought?' Arian says, smiling broadly.

'So, how's life for you in the big smoke?'

'Oh, Wren, it's just so enlivening in so many ways, but it's a terrible lifestyle — terrible! Look at my body,' he says, showing me a tiny bit of loose flab under his biceps. 'No muscles left at all. I tried going to the gym, but it was ghastly. It was like separating out in robotic sequence every section of a movement that I would naturally do with my timber work in the forest, and repeating it. Like a production line.'

The forest. There's a shared understanding of what we mean when we say those words.

'I was disorientated for quite a while. I came to realise that to really encounter the city, I had to come to terms with the immense amount of violence that's just everywhere.'

'What do you mean by violence?' I ask.

'Separation. Disconnection. One of first things I perceived was that the entire landscape is sliced and diced. What we're used to in the forest is this continuity of energy fields. It gradually changes as we move through it, but there are no beginnings and endings. But the city is like getting out a pineapple and a whole bunch of fruit and dicing it up into fruit salad, and all the bits are juxtaposed in incredibly intricate cubes. Rooms, houses, roads, cars, time itself, all the spaces are boundaried. Walking a street one day, it struck me it was an industrial drain — a passageway dominated by steel boxes on wheels for goods and objects and humans to move through, an industrial service conduit, for all the stuff that both feeds and is excreted by the house. *But*,' he says, sitting up in his chair and gesticulating with his hands, 'I realised the boundaries *also* enable twelve million people to live side by side in almost complete harmony. Just think about it — yes, there's crime, but there's billions of interactions going on in close proximity day after day, with very few problematic. It's amazing really! But what we need to remember is the field in which our bodies and imaginations are moving through, this landscape of relational poverty.'

Arian wipes his watering eyes. For a second I think he's crying. 'Pollution,' he says, 'often exceeds health standards.' Mine have been stinging too.

'But yet you're still here …'

'Awww, Wren, there's a single reason I'm here, and that's for the community. I'm learning about myself in a way I couldn't out bush. I've come to realise that what I came to London in search of was greater embedment. I was embedded in the forest in a way that was delightful and luscious, but I realised I needed my nervous system and body to be in interaction with both land and human ecologies.'

He re-crosses his legs and cups his mug in his hands. I can feel now how he's changed. It's like he's inhabiting his skin more; he's less diffuse, more rested in his seat. This being in the city has been good for him.

'So what is it exactly that you're learning?' I ask, bringing my knees up to my chest to try and stretch a little.

'What I'm exploring is new forms of tribe or community that are based on authenticity — ways to truly share our experience of the world. We're calling it "technologies of affection" — the skills and attitudes that enable us to enact this deeper realness with each other. It's all about pushing at the edges of this western concept of individualism, but instead of taking away personal freedom, we're exploring how to be individuals-in-relation: groups, networks, and communities who recognise we need to be moving beyond what separates us and find how we can stay anchored in all our uniqueness, and from there come into relation with others.'

'So rather than the old model of tightly adhered-to shared social values and norms, seeing how you can connect across difference?'

'Exactly,' says Arian emphatically. 'And what I find immensely exciting now is the shift away from framing authenticity as an individual endeavour and instead incorporating it into the current paradigm shift towards village again. There's a new understanding that personal authenticity enables a new sort of connection with others and thus enables new sorts of communities.'

'It certainly subverts the slicing and dicing and separations,' I say.

'And this is the reality of western culture at the moment — both incredible creativity and richness of personal freedoms within a field of disconnection and material violence on other humans and species. It's a juxtaposition every day for me, but yes, this is what I choose. My life is outrageously rich with a forest of humans, beyond my wildest dreams. That's why I'm here.'

'And I'm authentically celebrating your urban happiness,' I smile.

Later in the afternoon, I consult Google Maps to locate the nearest green patch. There's a largish swathe about two miles away. I need to stretch my legs and be with some trees after the long-haul flight. The parkland appears like an oasis in the concrete, its edges thick with nettle of a darker hue than that at home, as though it grows stronger when in its homeland. I find a grove of large oaks and lie down on the grass nearby and tuck my nose into the earth. It smells different, sweeter, less perfumed by astringent eucalypt. Smell is such a marker of place for me, even more so than sight.

I watch as several people nearby lean against a single oak. One checking their phone, one reading, and the other staring into the distance. I wonder if they're old or new acquaintances with this tree. The slicing and dicing feels far away from this peaceful continuity.

'Londoners love their trees,' the owner of a trendy bookshop told me when I wandered in on my way to the park. The shelf closest to the counter was dedicated to contemporary nature writing. I quizzed the petite older woman, dressed in a sky-blue skivvy. 'Oh yes,' she told me, 'nature writing is popular these last few years. Everyone is planting wildflowers in their backyards.'

I wonder if the oak is enjoying the attention as much as the visitors enjoy its company. I get up and walk on, pleased to find that the park gets wilder the further north I adventure, the grass growing long and weedy with patches of what feels like natural forest. There's something different that awakes in me in these places, away from cultivated gardens and mown lawns. It's something akin to being around a wild animal compared to a pet. This rumpled grove holds a quality that speaks of its own agency.

My senses are certainly not tracking for a large cat in the bushes, but I'm alert and curious. The wildish quality feels defined as much by where it is as by where it is not: rather than distant and pristine like the idea of wilderness, it's nearby and dishevelled, resilient precisely because of the persistence needed in the face of urban pressures. There's a pride to it — the self-contained sovereignty of a person who is authentically themselves in any company.

I've a newfound respect for the grit of these places after a conversation with Dave the ranger at the parklands where we hold

Rewild Fridays. Mel and I met with him with a certain nervousness, concerned that he was going to retract the permission he'd given us to have fires on our Friday programs during winter.

In half-buttoned khaki mechanic overalls and sporting long, straggly hair, Dave, launched into an hour-long diatribe on all the pressures he faced.

'The cyclists are organised and powerful. And don't get me started on dog walkers and their fur babies. You can't tell them about dog impact. Not my dog, they say. And then there's the bush kindergarten groups, multiple groups sometimes every day.'

I was trying to imagine how much damage a bunch of tots could do, but Dave read my mind.

'You know that grass that looks like pampas grass? We can't get rid of it with a shovel, but the kids have killed it.'

'How?' I asked, half-laughing.

Dave looked at me grimly. 'Mud. They're worse than goats.'

I shook my head in disbelief, picturing the army of mud-covered tots uprooting the indomitable grass.

'It's just not big enough,' Dave said. 'We have thirty thousand more people moving within two kilometres of here in the next five years. Since you've been using the area, paths have appeared around the wetland, and up the slope. Not saying it's you, but sensitive plants have disappeared. The already-shy swamp wallaby seems to have disappeared as dog walkers have begun to make use of the path.'

'Swamp wallabies?' How had I not seen evidence of them? In such a tiny patch?

'Oh yeah, they're there,' Dave said. 'Hiding.' Suddenly, his tone changed. 'But kids need this. I am where I am only because my parents grew me up amongst the trees. But collectively, all the groups make an impact, even the mums and bubs. It's about innocence,' he said with a sigh. 'I'm just trying to preserve some innocence.'

I felt a rush of gratitude for this man and the weight that he so clearly carried on his shoulders, doing his best day after day to both protect and promote these pockets of innocence, like the one I'm standing in now on the other side of the world. I move towards a hazel tree and, remembering the Yeats poem, snap a wand from a small dead branch. Looking into the heartwood of the hazel with the stick in my hand, I feel a bit like I'm in 'The Song of Wandering Aengus', alone in the wild woods and not in the middle of a metropolis. That's what these places offer, a reflection of our wholeness, our vitality and integrity. In their naturalness, they mirror our own essential purity, touched by culture on all sides but wild at our core.

It was a breaking of this innocence that shocked Jen, one of the more recent recruits to Rewild Fridays. After hearing of Laura's experience, Jen had decided to sleep a night in the group shelter. My first thought upon hearing her distressed voice on the end of the phone line was concern that she'd been hassled, but she quickly assured me this wasn't the case. Through tears, she told me how on a whim she had removed the small bird's nest within the thorny thicket that we had all been marvelling at on the day of the build.

'I just reached for it and took it like a kid wanting the shiny treasure,' she said.

It wasn't until she saw the eggs nestled in the downy tinder that the spell broke and she realised in horror what lay in her hands.

'The mother thornbill flew around my head, distressed. And then an egg fell out and cracked.'

Jen fell into sobs, feeling like she had been snapped out of an anthropocentric trance, a 'city spell of everything existing for human consumption or entertainment'.

'Isn't that what we all do here in the city?' I asked her. 'Steal life from the unborn?' It reminds me of Arian's comments about the violence of modern cosmopolitan life. Our city coma is well hidden under the complex chain of events, but it's not dissimilar. Stealing ancient sunlight to power our cars and computers, stealing viable soil and pollinating insects from the generations to come through our intensive crops. We just don't see the tangible results of our actions, the nests we unwittingly destroy, the death by a thousand cracks of our life-support systems.

'It's a strange moment in history to be living in, Jen,' I said, trying to normalise her experience. 'Witnessing our unconscious consumptive addictions. None of us are immune to that.'

Jen said how she had heard the distressed mother thornbill's muffled call on and off throughout the night from inside the pile of leaves, crying for its unborn babies. As a symbol of her commitment to be awake to her actions, she stayed awake to the sounds of its suffering, the sounds of the earth crying.

The story had an Edenic quality to it for me, the taking of the nest akin to Eve's taking of the apple. In this enduring cultural myth, humans are forever to blame for the destruction of innocence and beauty, cast out into a place of ugliness and woe. It's a paradox that, in its desire to claim responsibility, cultivates an anthropocentric separation.

Dave requested we move our fire across the creek to the established circle on the other side. It didn't have the same feeling of innocence, but I understood.

Wrapping my jumper around my hands, I pick a bunch of nettle to add to the evening's dinner and a posy of yarrow flowers for the table.

Walking the street back to Arian's, I keep thinking I recognise people to the point of calling out to them, before realising they're strangers. It's a thing that I do in foreign cities, my mind scanning for pattern matches like I do habitually with plants and birds. With a jolt of dopamine, I'm startled to recognise someone I saw on my walk *to* the park.

When I get home, Arian and his flatmates are celebrating the end of the workday in the German tradition of *fierabend* — a drink and a dance to wash off the day, perhaps one of the 'technologies of affection'. I accept a glass of bubbly and smile as I watch Arian chatting and laughing with his city tribe.

'See, this is me rewilding,' he jokes.

'I don't doubt that,' I smile, and it's true. Re-imagining community and human intimacy is at the heart of rewilding. A few generations ago, most likely anywhere on the planet, you would know every single person that you saw, and probably something of their story. It couldn't be more different now. As Sebastian Junger pointed out in his book *Tribe*, a person living in a modern city can, for the first time in history, go through an entire day, or even an entire life, mostly encountering complete strangers. They can be surrounded by others and yet feel deeply, dangerously alone. While we live in closer proximity, we are more socially distant — co-located but disconnected. Social isolation affects one in ten Australians, while one in six experience periods of emotional loneliness. The health impacts of chronic social isolation have the same implications for heart disease and stroke as smoking fifteen cigarettes a day, a UK study has found.

Charles Eisenstein is one of many contemporary philosophers naming the chief crisis of civilisation as a crisis of belonging stemming primarily from the dissolving of community. Modern technology and lifestyle, he says, have pushed us apart in a variety of ways and destroyed the stories that told us who we are and why we're here. 'People don't know who they are, because our identity is built from our relationships,' Eisenstein says. 'Without those relationships and stories, we have a hunger to belong, feel at home, and for identity.' Our innate evolutionary desire, he says, is to be interdependent, not independent.

Jon Young has translated some of the 'technologies of affection' of traditional cultures into practical suggestions, recommending that at the bare minimum we need to experience deep listening to our stories at least three times a day. How many of us actually have this even once a day?

'Loneliness does not come from having no people about one, but from being unable to communicate the things that seem important to oneself,' said Carl Jung.

My living arrangement is an intentional choice towards reimagining modern belonging. It's explicit in the way we choose to share food, gardens, and living spaces, and implicit in the ways we care for each other, and the way our social lives constellate around the home. Sharing a house is a choice we make, not just a product of economic necessity but a desire to live in community. There's a constant flow of people in and out the doors for gatherings and projects. We host potlucks and seasonal celebrations, working bees, parties, clothes swaps, and whatever else anyone has inspiration for. And it takes commitment — to communication, conflict resolution, flexibility, tolerance, and making time. Often, reality falls short of the ideals or expectations. We're in between stories, the experimental narrative of a new form of family co-existing alongside freedom and self-reliance.

Outside the house, I've nestled myself within various community niches with broad overlapping interests in embodiment, nature, human psychology, healing, sexuality, and creativity. It's a kind of loose ecosystem with clusters of affinity and belonging. We live, work, play, and relate together. The foundational shared value is community itself, self-referential in its purpose.

If technology has been part of the divide, it's also part of the reunion. Facebook is a valuable part of my toolkit. Our suburb has a potluck-dinner group, which extends into requests and offers for airport pick-ups, house-moving help, sharing excess produce, and spontaneous social adventures. The Rewild Friday crew regularly post observations, questions, and stories on our group page. Feedback is almost instant. This controversial commercial web allows me to organise, communicate, and connect with fluidity.

This is the reason I'm living urban. I came to know myself in the forest in the way only possible when reflected day after day by the untainted mirror of moonlight, river flow, granite stone, and whitening bone, stripping me back to reveal the unconditioned and uncivilised core of my being. Returning to the village, I've been putting the flesh back on my bones, exploring who I am via the mirror of the diverse human ecologies of the city. My adventurousness has found avenues in the relational, emotional, and vocational realms that provide opportunities equally as alluring and dangerous. It's another kind of experimental laboratory, bubbling with all the myriad ways we create, organise and arrange ourselves in the city. It's not the wildness of stalking the dark woods, but it's an equally fertile tension exploring authenticity and creative expression within the human hives and niches. I've come to locate myself in the way Arian has, through the affections and feedback of the human gaze.

Arian borrows a car, and we drive to Wales for a few days' camping in a patch of original woodland near the coast. Setting up camp, I'm like a kid remembering where everything goes. I open up the wooden crates. It's like the past has been snap-frozen in time: the same neat stackable Perspex plastic containers filled with dried fruit, spices, tea, sugar, nuts; the same teaspoons and mugs. I walk up to the old well to fill the wooden bucket, pausing to watch the light fall on the mossy rocks of the spring, which bubbles out from a magical opening in the forest bank. It's easy to imagine wayfarers over the centuries squatting to cup their hands at the natural fountain like me. I imagine what they might have been wearing, what their lives might have been like, if any of them could be my ancestors. A thin spire of smoke rises from our campsite, and I picture Arian blowing on the small flame. I slosh the bucket home and roll out my swag in my old place under the left eave. The sky darkens, and Arian hangs the billy from the hook. We've been quiet since arriving, perhaps stirred by the car conversation about our year together and all that has happened since. There's also a quiet warmth in the companionship of just being together in the woods, listening to its changing symphony as it slips into night.

Arian picks up one of the fire tools and gives the coals a thoughtful poke.

'You know, when I lived in the forest, the life force there was so powerful, a literal force field. It was confronting to realise how inherently fragile this physical form is and how nature is essentially indifferent,' he says. 'The opposite of love is not hate but indifference. It destroys ego and replaces it with a kind of realistic and humble understanding of your place in things.'

I nod, remembering well the shock to the ego at finding myself suddenly removed from all my usual social mirrors and placed before an animate world that had zero interest in shoring up my constructed sense of self. When the anchor began to pry itself loose, I was lost at sea, terrifyingly so at times — all ambition, all the bricks and mortar that had built the house of identity, washed away in the unrelenting tides of the forest. It was a cold-turkey approach to deculturation that continued well past the end of my year without matches.

'How much of our material hubris is a defence against this vulnerability?' Arian continues. 'How much of our obsession with human-centred environments is an avoidance of the bodily knowing of this truth?'

'That's what time in truly wild places offers,' I say. 'Shedding and letting go of our clinging to human identity.' Wild places certainly offer something even wildish city parks can't. When I returned from my immersion, the wild was tattooed on my soul; I was unshod at my core, and then from this place I could return to the human world and shape myself around that fundamental knowing. I think about Elena. I saw her right before I left and sensed the shift in gravity in her. The questions she's holding about her life have changed. She isn't looking to fix life anymore by tweaking her job or a relationship, but is ready to radically open up a conversation with the world. Her upcoming Vision Quest is part of that.

'In many ways, the forest taught me how to be in relationship, how to live,' says Arian, pouring the tea. 'I got to see how forest emerges out of each tree being fully and deeply itself every day and doing so in amongst all the other beings in its surroundings. Every tree's beauty arises from itself — sure, but that tree is growing into the space held by everyone around it. And every tree around it is doing the same thing. It's like an intricate dance that everyone is creating together. For me, this is how beauty arises. What we see as the beauty of nature is simply what happens when myriad individuals devote themselves to be as fully themselves as they can possibly be in the conditions created by those around them. And beauty also is fractal. In the forest, no matter what scale you look at, all the beings are doing exactly the same thing: from the tiniest bacteria and grains of soil, to the air, the clouds, and the stars, everyone is being as deeply themselves as hard as they can in the conditions created by those around them.'

It's interesting: authenticity has become the most common intention that the Rewild Friday crew speak at the beginning of the day as we circle up. There's a link emerging for me between the act of engaging with the beauty in nature that Arian speaks of and a desire for personal realism.

'Humans are more complex, of course,' says Arian. 'But the fundamental principle applies: being deeply and unerringly ourselves in the space held by those around us creates beauty. And the wildness inherent in that is twofold: we really don't know who we are or what we are capable of until we let ourselves find out.

'And what we can create together, right?'

'Exactly. We don't yet know our full beauty. And paradoxically, it's in the city — for me a place of great ugliness and violence — that I am getting a sense we might be able to find out.'

I rest back on the swag and look up at the first stars, hanging just above Arian's bald head.

'You still carry mountain man, you know,' I say, and he looks at me with a mixture of deep sadness and gratitude. 'Mountain man' was one of two figures that appeared to him in a kind of waking vision during his solitude.

'It's funny, a few people in London have said that being around me is like being around a mountain,' he says. 'And I've not shared the story here at all. It's hard to find anyone that would understand.'

'Yep, rare as obsidian,' I say.

'Well, you're a bit like obsidian,' he says. 'Dark and clear.' I smile and take his hand. We simultaneously raise our heads to look at the moon. It shimmers on, resilient and strong.

Chapter 9

'You look a bit like you've turned to stone,' I say to Min as she walks towards me out the front of the University of Melbourne science lab where she works as a casual geologist a couple of days a week. Her shoulders are unusually slumped. She rubs her hands up and down her face vigorously and releases a long and frustrated sigh.

'Shooting gamma rays at rocks all day will do that,' she says, blinking her bloodshot eyes open. Her half-smile is more of a grimace.

'Are you too tired to look at more?' I ask, pulling my hoodie over my head as a chilly late-afternoon wind picks up.

'No, never!' She wraps her scarf around her head and links her arm through mine. 'As long as it's fluorescent-light-free.'

I'm pleased our long-awaited geological walk of the city is still on. The timing is perfect. Even though we've been talking about it for months, it's only been the last few weeks that my interest in what lies beneath my feet has grown. My attention keeps being drawn downwards as if beholden to gravity — tracking the changing colours and textures of soil and sediment, running my fingers over exposed seams of sandstone, taking pleasure when rock rebelliously erupts through footpaths and formal gardens, like giant teeth piercing the groomed city skin.

This is a new curiosity for me. I suspect the allurement with rock has something to do with its apparent stillness and stability — an attraction of polarity to balance out the quickening pace of my mind. Time again has begun to feel like it's speeding up, carved up into bite-sized chunks to be eaten on the run. I need a way to place myself once more within earth's rhythm, and perhaps I'm looking to the rocks beneath my feet for a way in.

The sixty-four green acres of Carlton Gardens, home to Melbourne Museum and the Royal Exhibition Building, sit at the northern tip of the CBD. Min pulls up at a park bench and motions for me to come sit close to her. She draws her iPad out of her bag and clicks on a geology app.

'By the way, I loaded all the data for this app,' she says with a roll of her eyes. 'It means we should be able to access it without the internet. Fingers crossed.'

With a click, the grid-lined map of the CBD fades and transforms into a series of contour lines and bright colours in strange blob shapes.

'So here we are,' says Min, pointing to a large yellow area. 'It's what they call the Melbourne formation — siltstone and sandstone. From the Silurian era, laid down 420 million years ago.' I toss the number around in my mind as the wind whistles in my ears. It's hard to get any real feeling for the time scale. I pull my imagination back through history like a rubber band in an attempt to meet the numerical age of the ground upon which I stand, but it snaps back, unable to stretch to a time that long ago. I'm not sure how to make sense of such a number so it's not reduced to an abstraction or a conceptual factoid.

'Right now, we're sitting on top of the remains of a huge underwater landslide that rolled down the continental shelf. It's what they call a submarine fan.'

'Submarine fan?' I say in bemusement, enjoying the unusual phrase. 'Is this the type of rock I see near the house?' I ask.

'Yep, that's it. There's some sections where you can see the layers of sediment in a clear delineation of time and flow. The cascade of sediments falls down the slope and loses energy. Each rock and grain falls in a way predetermined by its size and speed.'

I imagine taking a slice through the layers of earth beneath me, like cutting a cake of silky silt.

'Let's get walking. It's damn cold, and I want to show you something,' Min says, tucking the iPad into her bag and pulling me up. The street is busy at the edge of the gardens, the crowd thickening as we approach the CBD.

'What are you noticing about the landscape?' Min asks as we wait for the walk sign at an intersection. I stop and look around me, wondering what Min is seeing that I'm not. I'm enjoying how Min's questions assume no distinction between built and natural environment. It's making me exercise my imagination in order to look beyond the moving lights and nameless faces to an overall lay of the land.

'Well, it's a gradual downhill,' I say, realising I've never noticed how the streetscape in the CBD channels downwards. 'It's like we're heading towards a creek.'

'Exactly,' Min says, flipping open the iPad again. Elizabeth Street is a wavy green tail on the colour chart, indicating a waterway.

'What they called Williams Creek is now a stormwater drain under our feet. Look, here's a photo of a distant cousin balanced precariously on a bench surrounded by floodwater back in 1972,' Mins says clicking on a photo — it's a veritable tidal wave rolling down the centre of town. We zigzag our way down the slope and pause at what feels like the lowest point, moving to one side as streams of peak-hour pedestrians flow past. I can almost imagine water lapping at my ankles. 'And this here was the start of a lake,' Min says, confirming my reading. 'With a small wander, we've crossed from yellow to red to green, a leap that takes us from 400-million-year-old deep-water sediments to 20-million-year-old basalts and then onto river flats — once lagoons and now rivers — that were laid down over the last eleven thousand years.'

I feel a little dizzy as I try to bend my head around the figures, as if I'm looking down into the abyss of time from a great height, albeit one still precariously poised on a single tick of the clock. I can sense the vastness beneath and above me but still don't know how to access it.

'Look,' Min says, pointing out dark fissures in the laid pavement stone. 'Basalt bubbles.' I look down at my feet to where the perfectly cut street stones are dotted with small bubbles.

I slow my walk to a stroll, Min matching my pace. Following the basalt bubbles on a direct contour downhill, I spread my gaze as I do when walking in the bush, the colour and movement merging together. I begin to feel like I'm walking with a foot in two worlds. One in the quickened footsteps of the peak-hour pedestrians intent on their next destination, and the other making tracks in the sandy riverbank at dusk, hearing not the honking of horns but that of the coots and marsh hens, breathing in the smell of a campfire, shocked by the sudden screech of sulphur-crested cockatoos flying in to roost in the tops of the river red gums. It's an overlay that's surprisingly easy to conjure, even after we reach the dead end of the major train station. The creek is still flowing somewhere beneath us, but we need to detour around the station in an attempt to find it again. We search, but the point at which it flows into the Yarra is invisible, paved over and ducted below the waterline. I feel sad that this penultimate place in the creek's journey is no longer witnessed. We instead sit on the concrete edge, dangling our feet out over the water.

'Can you imagine the wetland that was once here?' Min says with a sweep of her hand. I can. Cormorants perched on fallen branches with wings outstretched, yellow baubles of wattle reflected like small stars on the still surface, a spreading pink sky in the west flung through with flocks of corellas. I sink into the stillness of the scene, the reliability of its daily unfolding.

Min is hungry, and we backtrack to find a cafe with a window seat and order hot chips. 'If this was *Continuum*, it would be all neon lights and high density,' Min says, looking out onto the busy street. It's a reference to the dystopian TV series she's been hooked on. 'Strange to imagine the future when we have been imagining the past.'

'And what's it like for you, knowing this other reality beneath our feet?'

'It's comforting. I feel a sense of perspective and humility. We're just specks in the vastness of geological time, and this feels freeing in a way. The busyness of this city seems to lose its importance when you think in a scale of millions of years.' Her eyes suddenly glint with playfulness. 'You know, one of my favourite things ... I love it when the geology cracks through and the older story is visible, like when we passed the construction site and I just had to peek in to see the layers of time exposed. It reminds me that there is so much more going on than just the story of us humans.' She blows on a chip and looks out at the sea of people.

'And what if this chip could tell a story too,' I say, holding a fry aloft. 'Right now, I could choose to see earth's history reflected here in this potato.'

'Aha, like seeing it as a gift from hundreds of millions of years ago,' says Min, laughing.

Chewing on my evolutionary gift, I'm struck afresh with the limited experience of time that characterises our culture. It really is a spinning top compared to the graceful rotation of the earth, the slow grind of tectonic plates. Market forces blinker us into short-term thinking — success measured by economic quarters, a financial year. To maintain constant growth, the speed of activity must keep increasing. Political decisions extend as far as the next election —

four years at the most. Compare this to the Haudenosaunee people of North America, who consider important decisions in light of how it will affect seven generations hence. Our time horizon is too near, hiding long-term consequences.

Within this straitjacket of identification with the here and now, we see time as the enemy — racing against it, saving or spending it. Since time became money, our relationship to life has become impoverished.

Tom Brown Jr said that each time he went camping, he would spend the first twenty-four hours doing everything — setting up the tent, making the fire, eating — at quarter-speed, in order to slow down to what he called 'earth time'. I joked that I don't have time for such practices. But what if we could just shift as a matter of choice while doing the dishes or sitting on the train — widening our temporal context to include the larger story of past and future? Min and I just did, with our evolutionary potato chip. I had immediate access to a baseline far more restful, my lens widening to an experience of the moment that included both the swivel of the plastic stool and the deep-time perspective linked to the birth of photosynthesising plants.

Environmental activist Joanna Macy compares living in a shrunken bubble of time to agoraphobia in that it is self-reinforcing: people don't want to look too far into the future, because doing so can bring up despair and guilt at the legacy we will be leaving future generations. 'We can school ourselves to be aware, now and then, of the hosts of ancestral and future beings surrounding us like a cloud of witnesses.'

Swimming in an urban river, tracking the rocks of the city streets, foraging food from a laneway, making fire-by-friction in an urban parkland, I'm rubbing shoulders with the old and the new. The acts are more satisfying precisely because of the juxtaposition. There's a synergy there when I find myself located in both stories simultaneously — the youthful bustle of the city and the crone-like age of dirt itself.

'Hey, you were going to tell me about that whale house you went to last week,' Min says. 'You know, speaking of rocks and bones.'

'Oh yeah, that's right. It was definitely another time trapdoor.'

Marietta, a lively young woman who has been coming to Rewild Fridays, invited me to visit her aunt and uncle's house, where she was living. The invitation was based on the simple premise that 'you'd love it!' The clincher was the photos she posted on Facebook of some kind of pterodactyl nest high in a tree. She texted me the address with a note, 'You'll know which house it is by the whale bones.'

They were the first things I saw when pulling up: long crescents of grey-white bones hanging in suspended horizontal rows from the lower limb of a gum tree out the front of an otherwise nondescript suburban house in the civilised eastern suburbs. My gaze followed a spiral of spikes in the tree upwards to where a giant nest of sticks perched at the top, with a metal bat pole running back down from it. Despite the rain, I started climbing the rungs immediately, and was ready to descend via the pole when an aproned Aunt Betsie and ruddy-faced Uncle Rudy appeared below, waving warmly. I felt more like a street urchin than an adult dinner guest.

'Welcome to my favourite tree in the world,' Rudy called up, and I caught his European accent as I slid down the pole to greet them. They seemed unfazed about my mode of arrival, no doubt having witnessed many an adult-child climbing up the C. citriodora they planted forty years ago, when they bought the house and started converting it into their version of a cross between an equatorial village and a living museum.

'Come on in,' Betsie said and ushered me in the door. Inside, the walls and ceilings were stacked with artefacts, fossils, and curiosities: skull-shattering staffs from New Guinea, Aboriginal spears by the dozen, wooden masks fiercely adorned, shell and bone jewellery, hand-carved sailing equipment, and much more in the way of whale bones. It was as if the bat pole had transported us to a place and time that spanned continents and vast tracts of evolutionary time.

Rudy held out his hand to show me his first find: a palm-length arrow-shaped piece carved out of bone — a fertility symbol, found in the backwoods of the Netherlands, where he grew up. The next piece he brought out was a perfectly round shell bracelet. 'It would have been made by abrading the shell hour after hour, and then laid in the muck for thousands of years,' Rudy said. I put it over my wrist. It glowed as if pure moonlight.

Marietta was then instructed to hand me the 'mystery'. She carefully removed from the shelf what at first glance looked like a large white rock and placed it in my hands. My body tightened to contain its weight. I brought my face closer to the piece. It was smooth like marble. A set of white calcified patterns ran down one end, like a pen engraving.

'Animal, vegetable, or mineral?' Marietta prompted.

I closed my eyes and moved the object around in my hands. Without sight, my imaginative vision switched on. There was a deliberate structure to it, a reason why it was this particular shape. It had purpose once. A sentient life. I snapped my eyes open to Rudy's jubilant announcement that it was the tooth of one of the woolly mammoths who roamed the planet for about 250,000 years before the freeze came, vanishing from Siberia about 10,000 years ago.

I gasped as I felt afresh the weight of this prehistoric tooth. I could suddenly see the way the enamel had worn down over a lifetime of chewing herbaceous plants into a single white wavy line. The underside was rugged, the giant remaining roots of the teeth like stalactites hanging from the roof of a cave. I could picture the mammoth digging into the snow to ferret out the tundra underneath, piercing the ice on the edge of the lake. I could hear the sound of it bringing those four molars together to grind down the tough winter roots to make them digestible, the birds calling in the conifer trees above.

What had this animal seen in its lifetime? Could it have looked into the eye of one of our early ancestors? I suddenly had the odd feeling of being outside time, as if it was stretching like a long blanket around me. Aeons. The word echoed within me, the sound of it as ancient as its meaning. Perhaps it was whispered by one of the painted wooden masks hung on the wall around me.

'He found it at a market for not much more than a penny,' I say to Min, her whole being riveted to the story of this suburban prehistoric encounter.

As I stood there with the tooth, I had a clear sense of being shaped by the evolution of this strange creature, and all the beings who came before us. There was a time when nothing on earth breathed. Breathing had to be created. Every time I breathe, it's an opportunity for remembering the moment when life first drew breath — a memory passed down to us over millions of years. These cosmological experiences are available in the here and now of our humanness, even in the intelligence that has allowed us such rapid technological advances.

'It was electrifying, as if for that moment I really had stepped out of linear time and felt myself to be this tiny character in an aeons-old story,' I tell Min.

'Aeons and stories,' she mutters thoughtfully.

Betsie chimed in to tell me that some people cry when they hold the tooth. Cry, wow. Perhaps with an even deeper sense of the relief that I was feeling. The minute-by-minute change in estimated arrival time I had been tracking on Google Maps while driving there suddenly seemed absurd with this object in my hands. I didn't cry; I giggled. Contact with this weighty piece of history had lifted some weight of the everyday demands of my life. I wondered if instead of resisting this reality, I could hold it more lightly, wrapped within the larger circular tent of cyclic deep time.

Others around me seem to be making their own contacts. Last week, a regular mentoring client turned up looking harried and stressed. Kara spoke of the unrelenting busyness of her life and the slowness she craved. She poked a stick into the earth with resignation, and as if in answer, a blue tongue lizard appeared and slowly crept toward her. She froze as she watched its approach. 'Wow,' she whispered, with some tears. 'How beautiful. I've never seen one so close up before. It's so slow. So ancient-feeling, just like what I'm yearning for.'

Another young woman who sat with me spoke of how she balanced out the demands of her full-time job with her passion for hunting down in dusty library archives the Indigenous names of places in the city and surrounds. 'The old names, the old maps help connect me to a more mysterious sense of time and place,' she said.

At a workshop, I invited participants to write a letter to themselves from the future generations. Many returned having been instructed to spend more time in nature. I wasn't surprised. Nature is our best and most direct mentor for this alternative sense of time, revealing its cycles daily in the turning of the earth, monthly in the waxing and waning moon, and annually in the nuances of seasonal change. I rejoice every time I'm reminded of both how trackable and mysterious are these cyclical indicators, with linkages between seemingly disconnected ecologies — mullet and ironbark — wombat and orchid. My nature journal is still in its infancy. If I continue to take notice, more patterns will be revealed, and in the noticing, I'll embed myself within them.

'Speaking of time, our tram goes in ten minutes,' Min says, wiping the oil and salt from her fingers and standing up to leave.

As we head out into the fray, I'm already thinking about who else might wander with me in deep time.

A train rumbles overhead, and I lean into the side of the enormous brick pylon that I'm standing under. The park is empty, and I check my phone, wondering if I've got the right place. Enterprize Park, this is it. I wasn't quite expecting it to be a postage-stamp patch of grass sandwiched between river and train station. A man with a goatee wearing a small black daypack appears down near the edge of the park, taking photos. That must be Dean. He turns around as I start walking towards him, and gives me a nod.

'G'day,' he says, extending one hand in my direction with a broad smile, the other hand holding a takeaway coffee cup. Dean Stewart is a Wemba Wemba–Wergaia man who runs cultural tours in and around Melbourne and has offered to take me on a walk.

'Welcome to my office,' he says proudly, black curls protruding from under his cap. 'The place where the first white settlers came to Melbourne in 1836. This is where the cultural tsunami began.'

'This was the exact place of their landing?' I ask, my words drowned out in the rumble of another train passing metres over our head.

'Yep. Sure was,' says Dean with a slight grimace. 'Imagine us standing in waist-high grass. Behind us would be a wattle and eucalypt forest, emus and kangaroos moving through. Further back,' Dean says, pointing towards Flagstaff Gardens, 'was a primeval she-oak forest, and another to the south at what's now called Emerald Hill. Both would have been gathering and ceremony grounds.'

With this last suggestion, my imagination conjures a full sensory experience. The ring of she-oaks on the rise, the wind breathing the sweet whisper of the needle leaves down to the lowlands in invitation.

'And where they call Docklands, over there,' he says, his thick arm becoming signpost again, 'was a giant salt lake — a major migratory route for birds. Floods washed out this entire region. The tea-tree flats would have seen barren geese, black swans, swamp wallabies, wombats, bandicoots, spotted quolls.'

My imagination soars above the city. Weeding out the skyscrapers, I see the land exposed in all its wrinkles, bulges, and folds. Dean leads me around the construction fencing and down to the river's edge near the busy Queens Bridge.

'This bridge is where the river and ocean once met at a set of waterfalls made from giant basalt rocks, before settlers dynamited them,' Dean says. It's harder to imagine. He's a practised storyteller, though, inviting me to imagine the song of the waterfall changing with the tides. 'The river changes so frequently. Sometimes, it's just brown or so thick you could walk across it. And other times, it's virtually crystal clear, depending on the currents and the weather. Sometimes, all you can smell is the sea.'

Dean's tone softens as he begins to recount his own observations of the place.

'I've seen a dolphin here once. There's a seal that visits every year. There's snapper and barracuda. And see those gulls,' he says, pointing to a dozen or so birds gathered in the middle of the river. 'They're always in that spot. There must be an upwelling of some kind. I've only ever met one person, a birdo on a tour, who had also noticed that.'

I grow more intrigued as I realise Dean is speaking not just of historical belonging, but of contemporary experience.

Dean walks me over to the side of the river wall and points down. There's a small section of rocks to one side, a duck sitting on a piece of junk.

'You wouldn't know it, but these are the remnants of an important cultural and geological site — the only remaining rocks of the waterfall that separated river from ocean. These rocks are 1.2 million years old. There's a male and female swan here at certain times of the year with a gosling, teaching it how to eat the algae. I also love seeing the migratory welcome swallows here, nesting under the bridge. A pied currawong has come in the last eighteen months, and I've noticed it uses the alcoves of the buildings to amplify its song.'

'Really, you've seen it? That's amazing!'

'And listen to that,' he says, turning his head back across the river.

I pause and listen. 'You mean the rainbow lorikeets?'

Dean looks at me astonished. 'Well, you're one in a million. Seriously, barely one of the twelve thousand people I show around here every year can identify that sound when I point it out.' It's as if I've passed some test, and Dean grows more animated. 'Every individual who once lived here would have known the name of every plant and bird and animal, as well as the seasonal indicators. They knew it as intimately as you know your neighbourhood — the street names and houses. The elders used to say that it's written twice — once on the land and once in the stars.'

The sentiment fills me with wonder. Once on the land and once in the stars. How exquisite. It's poetry to me and I'm in awe of those who truly did read the patterns of nature via both land and sky.

'Traditionally, everything had a song and a dance and a story. Like the waterfall here. That's how things were remembered. Think about when a song comes on the radio that you haven't heard for fifteen years and the chorus just comes straight back to you. Songs stick.'

'And where do you hear the song of the land loudest?' I ask.

'Right here, mate!' he says emphatically. 'I know this place better than anywhere.'

It occurs to me that this man might know this piece of country smack bang in the middle of the city better than anyone, even though his particular lineage is from another part of the state. It gets me thinking again about what it means to be indigenous. I pluck up the courage to ask what feels like an edgy question.

'What does being indigenous mean to you?'

Dean adjusts his cap and looks out to the ducks on the river.

'At the end of every school tour, I ask that of the year-nine kids. They'll say things like "ancient" and "Aboriginal". And then I'll ask them to put their hand up if they were born in Melbourne and tell them: guess what, you're indigenous, you belong — and instantly their eyes open up and it's a gamechanger. I tell them that fifteen-year-olds have been walking on this landscape listening to the waterfall and the lorikeets and to an old guy talk about their connection to place for thousands of years. I tell them: you are the newest custodians of this place.'

A flock of silvereyes fly into the bushes next to us, and we both stop to listen and look. Dean turns around and gives me a wink.

'This is reconciliation in action for me, us as a people reconciling ourselves back to the land. We're all caretakers, we're all custodians.'

Chapter 10

It's dusk at the Riverhouse. Down by the riverbank, a single yellow-tailed black cockatoo screeches right above us. The six of us sitting in a ring around the fire simultaneously lift our heads. The bird is making a wide circular lap above the tree line. As if suddenly feeling watched, it falls silent, but continues to flap in slow, graceful waves, changing direction as if beholden to a sudden whim. The bird opens its thick black beak and releases a stronger, more urgent shriek. It could be the lonely song of a whale, a lament of a woman. The others have also fallen into their own quiet acknowledgement of its melancholic dusk expression. Years of alliance with the black cockatoo has suggested to me that this is a song that carries the voice of unmet longings, evokes yearnings too dangerous to speak. It's a carrier of mystery. A speaker of soul. The cockatoo lands on a high branch of the dead acacia in the neighbour's yard and looks to the west. One by one, we leave the bird to its reverie and return our gaze to the fire, where a dozen rocks are glowing iridescent orange from beneath the last layers of the pyre.

Cat rests her head back onto my lap, and I lay my hand on her hair. The soles of my feet are almost too hot, and I rub them together to cool off. Beth sits and stares into the fire with a peaceful intensity, the colour of her hair mirroring the flames. Chelle looks a little nervous, dusting her clothes of the falling ash. Kate shares her sheepskin rug with Phillip.

'Another ten minutes, I reckon,' says Phillip, getting up off the rug and slipping on his shoes. He squints as he approaches the fire with a long fork to consolidate the rocks. A coal the size of a golf ball rolls out of the fire towards me, and I push it back with a stick. I bring my fingers to my eyebrows, singed from my own turn tending the rocks.

The last of the Saturday boaters row past, chattering away like the straggling lorikeets finding their way to day's end. I'm glad to have a bit longer to rest, watch the fire die down, and think about my intention. I don't want to take this lightly. Sweat lodge, or earth lodge, is one of the oldest ceremonies in the world, with some form of it found on every continent. I've sat in half a dozen over the years. They're all different, all challenging in their own way. Crawling through the blanket flap of a door is like entering another land. You never quite know what you're going to get. But I know what can happen if I waltz into a lodge as if it's the gym sauna — the hubris seared away with humbling intensity.

I glance over at our construction. Covered in layers of blankets and a tarp, the lodge that will soon house the six of us in a circle appears as a dark-green mushroom sprouting by the neighbour's fence. I hope they're not planning on a backyard barbecue tonight. Phillip follows my gaze and gives me a raised eyebrow that reflects exactly what I'm thinking.

We've done well. This morning, the lodge was another one of my crazy ideas that Phillip got immediately excited about. We drove out to the middle section of the Yarra, just outside our sprawling metropolis, where the tea-tree grows thick, and harvested a rooftop of saplings for the frame. It took us the rest of the day to measure and sink the saplings into a circle, lash them together in the middle, and add horizontal rungs and an arched door. A light rain joined us for the last hour, as we tied the last of the lashings, dug the pit in the middle, laid tea-tree leaves out on the ground beneath the saplings for cushioning, and covered the frame in op-shop blankets. There's something about creating a circular structure that just feels so good. It's so radically unlike any other shape we usually occupy, and yet is more than symbolically satisfying. Just as a fire's gift is to warm, a circle is surely designed to gather within.

The idea for the sweat lodge came after I realised that the same basalt bubbles that Min had pointed out to me matched with the melon-sized sharp-edged black rocks that dotted the property. They're volcanic, I had thought excitedly — made for fire. A dozen were chosen and set in the centre of a log construction. As if it was indeed a funeral pyre, we gave the rocks a silent blessing, blew a tinder bundle into flames, and fed it to the kindling nest on the western side of the logs. The rocks were given new life in the flames, absorbing the heat with a hungry passion as if reunited with a lover.

The group starts stirring, readying themselves with water bottles and sarongs. My belly grumbles; I haven't eaten since breakfast, in preparation for the cleanse.

'Can I take a water bottle in?' asks Beth.

'Can I leave if want?' asks Chelle.

As Phillip gets chatty about protocols, Cat braids her long hair back into two plaits. I glance up the hill to where I can see the blue flashing lights projected onto the neighbour's lounge-room wall from their TV. The Saturday-night motorbike hoons have started up across the river. And here we are, about to enter into a practice thousands of years old. I like the edginess of it, the grittiness, the creative urgency that comes with finding ways to slip between and through the cracks.

At the last minute, we strip off and stand in a line outside the lodge, like six shivering white ghosts. The day is a mere red smudge on the western horizon, fading quickly to the city's shade of evening-time grey. I kneel at the entrance and touch my forehead to the ground. It's cold and clammy, smelling of mud and clay. There's some reluctance to enter, but I can feel Chelle's trembling impatience behind me. I crawl blindly into the dark until my hand makes contact with skin, and I settle into place next to Beth, both of us hugging our knees to our chests. The roof is barely tall enough to sit under, my hair getting caught on sapling twigs. I shuffle forward to find room. The others crawl in with similar quiet. We shiver with both anticipation and trepidation of the heat, as if waiting for life to enter.

Outside, the fire sizzles with disturbance as Phillip fishes with the fork for the first rock. There's a swish swish as wattle leaves dust the ash from the stone, and then heavy footfalls moving towards us.

'Heyo,' he calls in warning as the first rock enters the door on the fork and rolls into the bottom of the pit. As it settles, the stone spits small proud beads of fire in our direction. An immediate wave of warmth softens our goosepimples.

'Wow,' Cat whispers on the other side of me. 'It's almost ethereal.'

Another rock enters, and this time I guide it into place with two forked sticks. The clunking of their meeting suggests more emptiness than solidity, the rocks made of fire now more than of earth. Four more rocks enter, indignant sizzling orange sparks flying as they make contact. I nestle them together like eggs in a nest.

'Coming in, folks,' says Phillip as he too kneels at the door and pauses, before closing the two blanket flaps behind him. Suddenly, we're enveloped in complete blackness. Just to make sure, I raise my hand in front of me. I can't even make out its outline. The only thing visible is the cairn of red-hot rocks in the middle. The radiant heat is already intense on my shins. My underarms are rapidly moistening.

For a few minutes, nothing and nobody moves. There is silence, bar the soft breathing of Beth next to me and the hiss and crack of rocks cooling. Crickets suddenly pick up their night song, but it's more of an echo or memory, the sound muffled as if hearing it from down the end of a long tunnel. The stones demand my full attention. It's as if I'm looking in at a crater of small meteorites, still sizzling with the elemental power of their trajectory through the atmosphere, thrumming with wild energy and untamed strength. Someone offers white sage onto the rocks with a prayer of thanks, and the smoke enters my nostrils, sweet and acrid at the same time.

Phillip dips the ladle into the bucket of water. The sound of the drip of water is as loud as if dropping into the bottom of a well. The rocks hiss and spit with their first taste of this cooling element, and the clouds of steam rise up to the roof before falling down over us with surprising intensity. Phillip offers two more ladles, and already my pores are streaming with perspiration. My centre of gravity drops from head into body.

'Let's offer our gratitude, folks,' Phillip begins. Our voices are gravelly and low, but our words are like arrows. There's no oxygen or energy to spare for waffle. Gravity is a force pulling me deeper in and down, and my thanksgiving is spoken with one hand on the mud of the earth and one on the earth of my body. The heat activates the tea-tree like a humid summer day, and the lodge fills with its pungent, fresh scent as Phillip ladles more water on at the end of this first round. There is movement next to me, and Cat has laid her head behind me on the ground. I bow my head and close my eyes. I am filled with a sense of the elements present in this ceremony: earth in the wood that rubbed together to create the first spark and in the ground beneath us; fire that we nurtured from a small coal and that's present still in the glowing rocks; water from the river and from our bodies; the humid air we breathe. Giving thanks. Giving thanks. Breathing. Breathing.

The second round begins. The prayers continue, around and around the circle. We speak of prayers for others, human and otherwise, prayers for ourselves and the dreams we hold for our lives. There are tears across the circle from a voice in the dark. We breathe louder in solidarity. How long have we been in here for? The thoughts send a wave of fear through me and the desire to fall to the ground. If I try to resist this heat, if I even give in to the fear for a second, I will be crawling for the door. The songs begin. The lilt of voices creates a whirlpool. I open my throat and find that sound comes out. I'm swept up in song. The heat rolls over me in searing waves. My nostrils burn and my lungs swim. Everything around me turns to water — the air, the humans, the roof. Water drips over my lips and down my bare chest. I become a human river, melting, dissolving, softening. There is nothing solid to hold on to. All is fluid, all is movement. My lungs seem to fill with water, and for a moment I panic, thinking I've forgotten how to breathe. My heart beats faster. It's telling me that I need to fling the door open and escape. But I don't. I want to sit in this fire. To meet reality as it is, even when it burns.

I open my eyes and stare back into the epicentre. Suddenly, the age of the stones, the aeons-long story of the ground under me, hits me with an overwhelming sense of awe. This action of fire and stone is the longest story of time. Since the big bang, the earth's adventures have been primarily geological — volcanic flames and flowing lava and steaming rains washing over the shifting bones of the continents and into the shifting seas. Only so very recently has organic life emerged, the salt from the early seas still present here in the sweat from our bodies, in the tears of our prayers. What a tiny fraction of the story humans occupy. And this metropolis surrounding me for kilometres with all its millions of TV screens and fast cars — a mere microsecond, a notification arriving just before midnight.

I feel like I'm back in the earth's womb, swimming in a liminality of time and space. In the presence of earth, air, water, and fire, I witness the ordeal of matter refining itself, the alchemical process of elemental transformation. There's a restfulness to it, a steadiness. The knowledge gives me renewed strength, and I sit up tall and let the steam wash over me as I imagine it once did when lava met with the ocean.

'Last round,' announces Phillip, picking up the drum. Our voices rise in crescendo, singing for everything we have named or not named, singing in pure celebration of life, holding nothing back. I imagine our voices rippling out from the lodge and into the night, landing in the laps of the neighbour's watching TV, spilling out onto the street and moving through the city streets like a river of love.

'Door,' says Phillip, and I find physical strength to fling the blanket open. One by one, we crawl out and collapse onto the cool grass. My heart is pounding. I take giant gulps of the air. Next to me, Beth sighs long and deep. I open my eyes. Above me, the stars seem like they are hanging just above the canopy, so low, so bright, shining even in the city haze. It's as if I'm falling down into them rather than looking up. The leaves rustle above me, and the sound is of the softest greeting. I lift my head. Everything is familiar but pulsing with a luminescence, a pure life force, celebratory. I'm aware of the spaces between things more than the things themselves. The sound of crickets rushes up to greet me. I slowly peel myself off the grass and wander down to the water's edge. Five other ghostly bodies follow silently behind me. The surface is still, just a subtle swirling indicating the slow direction of travel. With a loud whoop, we dive in, the cold as intense as the heat. I break the surface shrieking with laughter.

Dawn the next morning, I walk back down to the river. The ashes from the fire are still warm. The rocks are not — stone cold in the pit. I lift them one by one and cradle them in my palms as I transport them to the fire circle. They are changed, faded to a dusty grey, and significantly lighter. As am I. The sweat lodge is again a tarp covering a stick frame, no longer the portal of the evening before. But when I look again into the bones of the fire, I can feel the bones of ancient history inside me, the story of my larger self, my earth body.

Chapter 11

We're all gonna fry. Can we push it back until the southerly arrives?

I message the three others on the 'kayakity yak' Facebook message thread. It's 4.00 pm. We were meant to have set sail hours ago for an overnight kayak mission into the CBD. But the temperature gauge is pushing forty-four degrees Celsius, and the river is hot as a highway.

I feel a little worried about making it too late, Daryl messages. *But let's be prepared as much as possible.*

Cat diffuses the situation by posting an animation of a red smiling sausage dancing on a surfboard. *No need to be the speedy hotdog.*

I peer out from my bedroom burrow, where I've been taking refuge like most creatures today. The heat shimmers like cellophane off hard surfaces. The pumpkin leaves in the garden below are crinkling up like brown potato chips. Even the dogged dogwood tree is releasing leaves from its grip: they float slowly to the ground as if in a dream. A grey fantail perches low in the privet outside my window, its beak open and panting. I brave the anticipated blast of hot air in my face to put a bowl of water outside near the bird. After a time, it hops down for a furtive drink and splash. I re-wet the scarf I've had tied around my waist and get back online to check in with my fellow pirates, who are similarly skulking in their nearby bunkers.

The urban river adventure has been in the dreaming since I floated the idea at a party over a year ago. The vague discussion of the last couple of months has solidified into three takers and a subsequent back-and-forth about plans and things to bring — water bladders, mozzie nets, gloves, water shoes, dry sacks for sleeping gear, maps and possible locations to sleep, zip-lock vacuum bags, and transport and rendezvous arrangements at the other end. The anticipation has been fun already, but no amount of planning can control the weather.

Like Daryl, I'm aware of our time constraints too. When we finally launch, we will be officially homeless, having no idea where we will rest our head. Our aim is to find a safe and dry place by nightfall as well as to have covered some of the twenty-two river kilometres that will take us to the heart of the city via water.

I've resisted suggestions to try and put a pin on a map to determine where we might sleep — I want to amplify rather than diminish the sense of uncertainty. I'd rather not take a map at all. The river flows downstream through the city and into the sea. Surely that's all we need to know?

We're still debating leaving times when I spy the tiniest of movements in the gum leaves. A mere flirtation, or a forerunner of the promised southerly? It's not meant to arrive for at least another hour. I step outside. A timid snake of sweet, cool relief wraps itself around my neck. Within minutes, the air temperature has dropped by five degrees. There's a creak of floorboards upstairs as catatonic housemates wrapped in wet cloth stir. A kookaburra calls, the clarion caller proclaiming the blessed arrival of the southerly.

Come on over, pirates!

Within the hour, we're sweating our kayaks and gear down to the neighbour's jetty, which, thank goodness, we now have permission to use again, having received a text a few months ago announcing that the fatwa had been lifted.

'Woo hoo, it's finally happening,' smiles Cat. I look around at the raggle-taggle bunch on the dock. With her mermaid-long flaming-red hair, fair skin, tall slim build, and broad-brimmed Akubra, Cat looks a bit like a plucky northern-European explorer heading into the interior.

'Group photo, everyone,' Daryl says, gesturing for us all to clump up in front of the kayaks. My experience of Daryl, a film-animation expert, has hitherto been limited to him being the owner of a suburban uber-mansion that hosts infamous hot-tub parties. With a tattered straw hat, and big hairy arms poking out of a sleeveless Hawaiian button-up top, Daryl could be in a film himself (less arid, more Californian coast). He's been right onto finer points of distances and rendezvous plans the last few weeks, and I'm glad to get to know his practical, adventurous side.

'Not quite ready,' says Oz, looking a little flustered as he fusses around with his gear, of which he has much. He also has by far the most advanced kayak. I've only met Oz once and know that he lives in another of the riverside rentals upstream. Unlike my throw-a-few-things-in-a-garbage-bag-this-morning packing approach, Oz has been posting photos on Facebook the last few days of his gear all laid out, with brand-new dry bags to hold it all together. He brandishes a large sack of energy bars.

'That'll get us to Tasmania,' I laugh as he stuffs them into the hold.

'Well, while we're waiting,' I say, and strip off to my underwear and dive into the river with a giant splash.

'Yeah!' says Cat, jumping in behind me, followed by Daryl.

'Come on, Oz, it's our launch-off celebration,' Cat calls, and Oz jumps in with a holler.

'All right, let's go, we've still got to find somewhere to sleep tonight,' Daryl reminds us, clambering back onto the jetty. Despite the southerly, it's still hot enough for my hair to almost dry by the time it takes me to get my vessel packed and in the water.

Daryl lowers himself into the kayak. It's a snug fit, and the water level rises almost to the rim. Oz is the last to set sail, stuffing the last of his gear into the hold.

Min and Megan race down to the jetty to capture our departure on film.

'Bon voyage!' Min calls.

My first paddle stroke propels me effortlessly away from the jetty. The next feels even more smooth, as if I'm moving through silk. A sense of enormous freedom washes through me as my paddle slices through the water.

'Yaaaaa hoooooo … !' I call back to my flatmates, who are still waving. I feel slightly self-conscious knowing I'm on film but elated to be finally moving. The four of us look around at each other and start laughing.

'We're doing it!' Oz says to no one in particular.

The first sharp bend directs us away from home and towards the first stretch that's empty of human signs. Enveloped by green, the river red gums are like a parade of well-wishers launching ticker tape of sickle-shaped leaves down on us. A single white cockatoo perches within centimetres of the water, looking stunned, barely raising its eyes to our entourage. I can hear the bat colony up ahead, albeit quieter than usual in the heat.

We float under the Eastern Freeway bridge, the sound of six lanes of traffic a harsh juxtaposition to the gentle drip drip of my paddle. If I was a car up there speeding towards the city, I would reach my destination in about twenty minutes; via this twisty, waterborne, non-motorised route, it will probably take about eight hours of straight paddling.

Even though I've only dropped a couple of metres from riverbank to water level, I feel like I've entered an entirely different reality. My kayak rocks back and forth with the momentum of my paddle. I can feel the current and the way the water moves around me and with me. Droplets slide down the paddle and onto my thighs. It's so unlike the bubble of a car, where I'm buffered and alienated from the landscape I move through; here on the river, I am in it — travel route and travel medium synonymous. We fall into an easy rhythm, alone but together, the damp sarong around my neck netting the breeze like an air conditioner.

I look around to see Cat trailing far behind. I drop back to check on her. She looks grim.

'Yeah, I'm tired already. And we've only just started.'

'Really?' I say, surprised. Maybe I've got an easier kayak. I feel like I've just been drifting downstream.

'Look,' she says, pointing. I look around to see a grey-headed flying fox swooping in front of our boats, low enough to touch the water before finding a roost in a low shrub. I bring my binoculars up to my eyes. As the bat hangs upside down, I watch the water from its belly dribble into its mouth.

'So that's how they drink!' I exclaim.

Cat nods, knowing a bit about bats since taking the kids she works with regularly to the flying-fox colony that we're approaching. Pretty soon, we're surrounded by screeching, swooping furry beasts. I have heard that they're dying in the heat and half-expected to see floating carcasses. Instead, they have left the exposed high roosts to find cooler quarters in the low shrubbery, affording us close encounters. Some flap their wings around them as if creating a fan.

'Wow,' Oz cries as one swoops within a metre of his face. With his camo kayak and look of jungle amazement, he reminds me a bit of a fresh-faced Indiana Jones.

'Hey, look — babies,' Cat says, pointing out a group of pint-sized bats hanging on to their mothers.

'Wow, you can really smell their piss,' Daryl says.

'It's actually their scent glands — their pheromones.'

I hang back as the others forge onwards, wanting a little time alone here. This is the furthest I've ventured downstream via boat. I'm loving not knowing what's around the bend. Curiosity is such an inherently pleasurable feeling. I balance my paddle and trail both hands in the water. The ripples from my hands run like skipping stones. *Here we are.* A languid thought floats in. *Here I am.* I splash my bare legs, the rivulets running down my ankles. The rest of the river stretches out before me in glorious unfamiliarity. I'm excited. This is the kind of time I've been wanting to share with the river since I moved in. This trip is not just about fun and adventure, but a way to encounter the river with greater intimacy, according to its terms and time-scale, its rhythm and flow — a chance to get to know its changing shape and personality, and witness how it matures beyond the bend I inhabit. Although I live in the 'Riverhouse', I've never really steeped myself in its world for longer than an hour or two. I'm an observer and admirer, but can I really say I belong to the river? That I dwell on the river?

That's why I decided to take the long and winding route to the city. It's an invitation to surrender to the curvaceous and twisting meander that a river makes on its way to the sea. I want to know by the ache in my shoulders and the blisters on my palms what it takes to travel to my destination via this most original and natural of travel routes. There's no rushing or shortcuts here. Just the next bend, and the next.

In the era of the internet and cheap airfares, it's easy to assume that one's identity is independent of place, that I could fly to a city on the other side of the world and experience the same thoughts or feelings as here. I don't want this to be my reality. I don't want my psyche to be separate from or empty of place. But it's entirely possible — if I don't ingratiate myself to a location; if I don't get low enough and immerse myself long enough to encounter the gifts and challenges, paradoxes and complexities of a place. Just as the river has shaped the city, this trip is a way of laying myself open to being shaped by the river too. I'm not yet sure what my final shape may be, but I'm here to find out.

The bats serenade us right up to Studley Boathouse. The tearoom and boat hire are completely deserted in the late afternoon. We tie our kayaks up on its dock and climb aboard for a break. A large sign warns us in no uncertain terms, 'No swimming, ever, at any point.' It's the kind of invitation Daryl lives for. Within minutes, he has his entire kit off and hands Cat his phone to capture a pontoon pose before launching himself with an enormous splash into the waters. The ducks are clearly accustomed to more-civilised, bread-offering human behaviour and make a collective indignant, quacking exit. Cat lines up next, Daryl capturing her in slowmo on his phone as she swan-dives in.

I'm a bit annoyed at the constant phone presence. With great relief, I left mine in my desk drawer, and the feeling of freedom that I am basking in has a lot to do with that decision. I'm not only blissfully unaware but also blissfully incapable of finding out about news, social events, or even the time unless I ask, which I have no desire to do.

'You've gotta watch this,' says Daryl, beckoning me over. I peer reluctantly over his shoulder to see a slowmo of Cat leaping out over the water in full suspension like a white puma. It's such amazing footage that I get him to replay.

'My turn!' I say and strip off, lining up at the water's edge for my own filmed swan dive. I enter the water like a spear, fingers first, making a straight line for the depths, where I break into some wide breaststrokes into the cooler layers. Wow, I'm not even bracing against the water, worried about its contents. Maybe it's something to do with the risk element of the mission itself — spilling over into greater risk resilience in general.

'Wild swimming right here, people,' says Cat. It's a reference to the rewilding movement's promotion of swimming in natural waters. Proponents blog about the transcendent high they receive from dipping into seldom-swum and often-cold rivers, ponds, streams, and oceans. We're taking it a step further with our skin-to-skin contact with the river. It's the less inhibited end of what's become known as 'blue mind' — the therapeutic benefits of spending time in or around water. Research has shown that even sitting by a fountain can be quite restorative.

Oz offers around his Santa sack full of energy bars. We stand around and munch quietly, taking it all in — the novelty of our floating transport, the late-afternoon light through the leaves, the river drying on our skin.

'We're really here now,' Daryl says sagely. I nod in agreement. I feel it too. Pedestrian time feels far away. We've entered river time.

'And unless we get going, we're going to be navigating rapids in the dark,' says Oz, packing his goodies away.

Around a few bends, we encounter Dights Falls. A swirl of dark-chocolate water joins us, and I shrink my knees back from the paddle drips that I have hitherto been enjoying. Since the Yarra joins Merri Creek in collecting a swathe of northern Melbourne's stormwater, it's illegal to swim downstream of this juncture. Although polluted, the appearance of this junction brings to mind the cast of faces of those I know who live along the banks of Merri Creek and love it, albeit from a look-don't-touch place, speaking of it as affectionately as if it was a household member.

'Land ahoy,' says Daryl as he pulls up shy of the two-metre drop that is the falls. A fashion photo shoot is underway at our pull-up point, and they pause to allow us passage. A tent is half-hidden in the bushes. The makeshift home looks recent, and I wonder who might dwell there and what relationship they might have to the river. The juxtaposition of the glamorous models next door to the tent captures my imagination. Water is a great equaliser: rich and poor alike are drawn to the water's edge. It's always where the village meets. We clamber over rocks to assess the stretch ahead, which features a small section of rapids. Although they're relatively moderate, Cat and I play it safe and carry our boats past them, mooring on some rocks downstream so as to witness the other two negotiate the rapids.

'Yeeeeee-haaaaaaa!' Daryl yells as he pushes off into the centre of the white water. After a wobbly start, he finds his balance and emerges dry, with a loud hoot and holler. From somewhere in the nearby apartments comes the sound of clapping. As Daryl fiddles with his phone in readiness to capture Oz's descent, he is unaware of his boat's slow slide towards a smaller row of rapids. He looks around just in time to realise he is fast heading for a collision with the rocks.

'Faaarkkk,' he says, and his grin turns to shock and panic as he tries to stuff his phone back into his dry bag. His boat hits the rocks and lurches sideways. Cat looks over at me, and we dissolve into paroxysms of giggles. As Cat loses it, so does her boat, which also runs aground, taking on a bucketload of water. She screams half-hysterically, her arms and legs flailing. I leak a teaspoon of the pee I've been holding onto, which only makes me laugh harder, and my boat also lurches sideways. I finally find clear passage, my laughter rippling downstream, much to the bemusement of the river power-walkers.

It's dinnertime, and we pull up on the muddy bank near Abbotsford Convent, which houses the volunteer-run donation-based cafe Lentil As Anything. It's a Melbourne institution, and we're all well versed in the cheap and cheerful fare we take out to the long wooden benches next to the Convent's extensive gardens. The novelty of our river arrival to this familiar gathering place is not lost on us, and we walk up to our dinner date with paddles in hand, laughing. Cat's flatmate Bonnie and visiting friend Lauren are there waiting for us, eager to hear the tales thus far. I'm resistant, wanting to stay in the immediacy of the experience rather than make it a story of the past. Their enthusiasm echoes the response that flooded forth following Daryl's Facebook photo post of us all sitting on a kayak in his backyard a month ago. 'Take me too!' the comments lamented. An entire flotilla could have been assembled if we'd said yes to everyone who pleaded to come along.

Between the lines, I read a clear yearning for adventure — and a disbelief that you can just, well, go on one, right here. Maybe that's why the FOMO was so strong, because we were creating a voyage here *in the city* rather than planning a horse ride across the Mongolian steppes. My inner adventurer lets me know very loudly if she's starving — starts clawing at the walls and getting grumpy with everything that represents modern life's manufactured order and routine. She wants, no *needs*, to be let loose in an unfamiliar landscape that includes uncertainty, effort, challenge, and the smallest hint of danger. Trying a new cafe just doesn't do it. I need to get my feet wet.

I've been experimenting with mini-adventures close to home. When my friend Mahli visited, we designed our own twenty-four-hour exploration — no phones, no entering a house, no food except what we forage. We camped in my backyard and kayaked upstream until the currents turned us around. Pulling over at an inviting shoreline, we napped on the grass, ate nobbles of wattle sap, climbed a tree, and foraged asparagus from a forgotten garden. We chatted in swags under the stars and woke warm-hearted to greet the sunrise. The adventurer sighed in satisfaction. Even climbing a new tree or wandering to an unfamiliar area in a park scratches the itch. These off-map explorations are small but medicinally strong antidotes to lives increasingly mapped out, raising the tolerance and appetite for uncertainty. Life itself is one big adventure if we choose it to be.

The light suddenly drops, and I'm aware of the need to find a place to sleep. We slip back into the river and start looking out for a place to bed down. Downriver is a green grassy flat with a makeshift bar, complete with shabby-chic wooden stools and a sign naming it 'Worn Again'. The flat is a perfect campground but clearly private land. We decide instead to pull up on the edge of a park, where we discover a grassy ledge hidden from sight just above the water line with room enough for us all. We begin setting up a basic camp of mozzie nets and sleeping bags.

Across the river, a stormwater drain empties a constant stream of water like an urban bubbling brook. As I fiddle with my mozzie net, it occurs to me that there hasn't been any rain for a while. The drain must be a permanent creek that's been concreted and channelled underground. I pause and try to imagine it as the tree-lined creek it surely once was. I get Daryl to check on Google Maps — it's unmarked. What name might it have been given by the first peoples of this land? What stories might it have collected? How many other creeks are now paved over and dribbling out somewhere without acknowledgement? We watch the last light fall on the river to the sound of our brook and a second round of belly laughs as we replay our near misses at the rapids.

Cat and I snuggle up under the same mozzie net and are just drifting off when we realise that ants are crawling into our sleeping bags. We're right on a nest. With much groaning, we haul our camp up to the edge of the open parkland under a dead but still-standing tree, now in full view of any park visitors. As I tuck myself in to my sleeping bag, I wonder what kind of nocturnal activity occurs here. Despite the question, I fall asleep as I always do when sleeping on the earth — quickly and deeply.

In the night, I wake from a dream to a rhythmic whoo whoo and open my eyes to the silhouette of a tawny frogmouth in the tree above. Its mate is returning the call in the distance. The day-old full moon rises and hangs like a tiger's narrowing eye above the feathered creature. Gosh, how I miss sleeping outside. Being indoors night after night is a theft of the magic of these moments, robbing both me and the dark other of these quiet exchanges. It reminds me that my mum said she would love to sleep under the stars, having never done it before. We talked of camping out together on the riverbank of the farm I grew up on, where she still lives. The river there is similar in size to the Yarra. My brothers and I would spend hours down there as kids, swimming, fishing, crawling through the stands of bamboo. There were always adventures to be had. As much as I love the ocean, it's rivers that keep drawing me to their banks. This summer, I promise myself, I shall take my mother to sleep by the river and under the stars.

I'm up and out of my sleeping bag with the first dog walkers. Cat sleeps on, her long form ensconced in a sky-blue sleeping bag like a giant chrysalis. I stretch nearby and observe the bemused reaction of the joggers, who keep a wide berth around the bedded caterpillar.

'Tea!' I call out as my little Trangia stove bubbles to a boil. Daryl appears, yawning like a bear, and greets me with a morning hug. Oz has already packed up his gear and bounds over with a smile.

'Sleep well?' I ask.

'Not fantastically, but I feel great.' He squeezes condensed milk into his mug. I take my first sip of sweet, milky Earl Grey tea. It's beyond good. Oz and I sip quietly next to each other, looking out to the river in a quiet camaraderie. There's something so refreshing about this gradual and natural gathering for the morning cup of tea, without the need for extended planning and back and forth messaging.

A few early kayakers wave and smile at the sight of our hobo camp. The others have started chatting away, and I have an impulse to take the last of my tea to a spontaneous sit spot.

The river feels somehow closer this morning, like I can *feel* its personality. An overnight is always a good start, something happening in the unconscious hours that lends itself to attunement and relationship. A pied cormorant flies in and spreads its wings on a fallen branch. The current is almost obscured by the ripples created by small drifts of wind scuttling the surface in the opposite direction. A leggy spider near the shore rides on the micro-waves. It's like I'm seeing the river without so much of a personal filter this morning: it's not just a carrier or passive recipient, but an active agent itself, alive with its own mythos, separate to my ideas or projections.

We're river pilgrims, I think to myself. It might be brief, but in a small way our river journey contains all of the characteristics of a pilgrimage. And like other pilgrims, I'm vulnerable to my limitations. Not in control of my environment, I open myself to numinosity, generosity, synchronicity.

My friend Maya Ward embarked on a pilgrimage of signifi-cance along this very river in 2003 — a twenty-one-day walking pilgrimage from mouth to source — and wrote about it in her book *The Comfort of Water*. She lived for a time on the Yarra's banks, and what started out as an idea gained momentum until it became an act that she writes as being 'at once utterly obvious and com-pletely necessary'.

Maya now lives again on the banks of the Birrarung, in a village upstream. She recently took me for a stroll along a river path that was unfamiliar to me and spoke of her great walk. 'The notion of walking the length of the Yarra grew from my quest to live with clarity and sanity in the place I call home,' she said. While preparing, Maya and her three compatriots realised they were planning the walk exactly 200 years since the acting chief surveyor of New South Wales first rowed up the Yarra during his survey of the place known to the local Indigenous people as Neerim. He named the river 'The Freshwater', describing it as the most beautiful he had seen.

Maya delved deeper into the Indigenous history. That included long conversations with Wurundjeri elder Ian Hunter, with whom she had worked with for many years, running community and educational events at CERES Community Environment Park.

'Ian wanted me to know that the whole river is a songline. And that made so much sense. The river is the binding thread for people here and has been for tens of thousands of years. He asked that the spirits of the land walk with us.'

Maya and her fellow walkers contacted the private landholders, many of whom surprised her with their rousing support and enthusiastic gratitude, as if her act of walking could carry their love of the river too, just as pilgrims on the Camino trail in Spain carry prayers on behalf of others. However, it wasn't until Maya stepped out the front door on the first morning of her pilgrimage that she realised what she was actually doing.

'To have lived near the river all my life and then commit to knowing it in a ritualistic way — as soon as I opened the door, it was like I was entering a dream … a field of intelligence that was conversational and communicative.' I watched her eyes moisten. 'Standing on my welcome mat outside the front door, about to set off, there was a sense of something turning inside out, and instead of being welcomed to one small house, I was being welcomed to the world.'

That phrase encapsulates the similar feeling I had during those first paddle strokes that took me away from home. It wasn't just recreation or exercise — there was intention in those strokes, in the desire to meet and be met by the river.

The echo of that welcome rang on for the entire twenty-one days, shattering Maya out of ordinary reality. 'Part of the reason it was so powerful is that the river was already a symbol of hope for me. Rivers, connecting all the land of the city in their catchment, also connect people in their love for them. Sharing my story of the river pilgrimage was powerful, as I came to realise the depth of connection people have to their waterways. And on top of this, there were years of powerful experiences working alongside Ian, being introduced to Wurundjeri stories and Indigenous thinking. When I began my Yarra pilgrimage, a field of meaning washed over me and I was inside it.' She had a sense of 'mythic time', and wondered whether it was a glimpse into an Indigenous experience of the world. 'The way I understand it, Indigenous cultures work with trance and altered states deliberately and skilfully, as they knew how efficacious they were. Hunting, for instance, requires intense stamina and openness to country in order to attune to prey. They needed to live much of their life in that state, that's partly why there's so much ritual and ceremony. You're more perceiving, you're in communication with your food, with your environment, you're *in* it.'

On a whim, I dip my toes in the water and swirl them about, feeling into this idea of the river as a songline, in the way Ian Hunter had described it to Maya, as 'a path of dreaming, a narrative of the ancestors, which mapped the land in song'.

'I remember, and it brings me to tears all these years later,' said Maya. 'Lying in the sphagnum moss and looking up and seeing the rain and knowing the winds come from the west and the waters are picked up in the ocean and they travel all the way to the mountains and there the rivers run down again. I felt the cycle of the bay.

If we're Melburnians, 70 per cent of our body is river water. You could think of us as a piece of river with hands and tongues, speaking about the self.

'When I got to the end, I had a sense of something, something that is impossible to name, but the word "ancestors" is important here. A sense of ancient things, ancient yet alive things, and somehow I was involved with it all, connected, in mysterious ways. It shook me deeply. I didn't know how to be with such a changed sense of things. For months after the pilgrimage, I started shaking every time I spoke about it. As a white person, what did it mean to have such a visceral sense of a web of meaning?

'Coming home was very confusing. I had so many questions. First off, perhaps, was what right did I have to deep and ancient meaning? I couldn't answer that. But over time, it became something like — what responsibilities do I have to culture and country now that I have been gifted this rich sense of meaning? These are the sorts of problems that we as bearers of the legacy of colonisation need to investigate. It's not an easy thing. It was over a decade ago, and I feel like I'm just starting to understand that experience as an innate part of the wildness that we all are and that we need to find ethical ways to reclaim, ways that honour the brutal past and are gentle with our animal selves.'

Returning from my sit spot, I join the others to forage juicy, fat blackberries for breakfast. As we sit and eat in our kayaks, I chew on Maya's words, on this difficulty in claiming our own relationship to place that both acknowledges and draws on wisdom from Indigenous history and is open to an individual and contemporary depth of experience. What would it be like to create modern pilgrimages for the purpose of re-mythologising our relationship to place? What if, before circumambulating a mountain in the Himalayas, we chose first to travel by foot the rivers and creeks of our towns? To make the journey from kitchen tap to the first trickle high in the mountains?

These waters are part of this city's dreaming. More than any other natural feature, they contain the carrying capacity for a modern myth that can weave us into the narrative of place. The river is a wild creature. It has its own agency, is subject to forces beyond my control. This small river pilgrimage is one way to embed myself into a new urban myth that meshes the civilised and the wild, that offers a doorway into a deeper connection to the wilds both within and out. With this possibility in mind, it becomes, as Maya suggested, an act of 'sacred geography'.

I've heard other stories of sacred geography. A man called John emailed me to share his own humble pilgrimage. 'I have tracked Merlynston Creek from its source at National Boulevard Nature Reserve to its confluence with the Merri Creek, a distance of eleven kilometres,' he wrote like a true explorer. His statement of intent was matter-of-fact and true: 'Water is life, and to understand watercourses gives you a profound connection to the natural landscape.' Though he wrote at length about the features of the creek, his story spoke most clearly in the one hundred photos he attached of his walk. But then he concluded: 'Yes, Merlynston Creek is now surrounded by urban sprawl and industrial development. But it also continues to provide some natural habitat and niches for indigenous plants and creatures. We have massively transformed the creek to be, at various stages, a drain, a flood retarding basin, a lake reserve, an underground culvert, and a cemetery feature to wash away our sorrows and memories of ancestors, and it still survives with integrity.' Another contemporary pilgrim willing to love the land he lives on despite its flaws — or maybe more so because of them.

The next leg of the journey takes us into the land of grammar schools and canoe clubs and *Great Gatsby*-esque riverside mansions. We pass under overhead bridges signposted with familiar road names. There's a cognitive dissonance to the very different experience of place down here at river level.

Cat is sore and tired and pulls over for a break at an embankment. We scout a thin piece of bush to pee within, then sit on a park bench to munch corn chips. An older woman sporting a crisp pink skirt and matching hat stops in front of us.

'You weren't the same kayakers who were at Dights Falls last night?'

'Yeah, that was us,' Cat replies.

'What, did you travel back again this morning?'

'No, we slept in the park,' Daryl offers somewhat proudly.

The woman looks excited. 'I saw you — I was at a friend's apartment, and we clapped you as you went down the rapids.'

'No way, that was you?' Cat says incredulously.

'Yes, it really was,' she says excitedly. 'I just happened to be walking along now with another friend and recognised your boats.'

I look up at her, suddenly very interested.

She adjusts her pink hat and looks out to the river. 'You know, I used to be a kayaker. I won a championship once. I loved it, the feeling of moving down the river. Such a feeling of calm, like I wasn't even in the city, but loved knowing that I was too, if that makes sense.'

'Yeah, I get it,' I say, nodding emphatically.

'But now all I have is a paddle,' she says, shrugging her shoulders.

'Well, I have a kayak, so it's not too late,' I say, hoping but not expecting she'll take me up on my offer.

'No, I'm too old now,' she says not too convincingly before bidding us farewell.

'Wow — that's ridiculous,' Cat says, looking around at me.

'Why is it that these kinds of coincidences only happen when you're travelling?' Oz says.

'Well, since we're travelling so close to home, surely that means we could travel at home too?' I say. 'Isn't it more a mindset rather than a distance from home or destination?' It's a question I've wondered about before — whether synchronicities happen more often when travelling or you're just more open to noticing them when the wheel isn't locked in the routine rut of awareness.

'I reckon I could travel in the backyard,' says Cat.

After a stop at the free-climbing range under the freeway at Burnley, and another at Herring Island Sculpture Park, we turn for the city. The river here is stripped of riparian vegetation, naked and exposed. The freeway hugs the rivercourse. I catch a sign indicating that its four kilometres to the city and wonder how far that means for us. We fall into a quiet reverie. Soon the skyline is visible and the green banks of the public parkland. The smell of barbecues floats along with us, and I remember that it's the national holiday of Australia Day.

'Invasion Day,' says Daryl, catching up to us and reminding us that our final destination — Batman Park — was named after one of the first settlers, John Batman. 'Where's our Aboriginal flags?'

The sun is beating down, and Oz erects a red-striped umbrella to the back of his kayak to shade under. How on earth did he stash that? We hit commercial-cruise land; the waters are thick with vessels, and the banks are packed with picnickers, Australian flags, and the smell of sausages and beer. Here, today, the river is a place for some to gather in celebration, but both the political context and the noise has me quiet and pensive. A rowing boat with a dozen synchronised oars careens towards me like a motorised ant. I narrowly miss it, one oar hitting my kayak. 'Get over the other side,' one rower yells in river rage. I look over to Daryl.

'International river rules,' he says. 'Keep to the right.' Good to know.

River cruisers pass us, waving and pointing. Daryl attracts more than his share of laughter and stares, especially now with his shirt unbuttoned to reveal a hairy belly. With that and Oz's sun umbrella, we look like a bunch of country hicks. It's almost impossible to find somewhere to moor — all the docks are reserved for the packed tourist cruises. We illegally tie up on some rocks and clamber up the slippery bank to meet friends at the River Bar, again with paddles in hand. After a few ciders, we continue the last leg of the journey, paddling a few hundred metres to our destination, where our return driver is waiting for us.

It takes the predicted twenty minutes to get home. It's too quick. I'm back before my land legs are. With hugs, I say goodbye to my comrades and drag my kayak inside the front gate. I stand near the front door and can't bring myself to step inside. It somehow feels disloyal to the river. I'm not ready to finish this pilgrimage yet. I walk back down to the jetty, where we started. It feels like days since we were here. I dangle my hands in the water and journey back through my memory — the places seen and visited, the bats, the birds, the groves, the blackberries, the silence, the camaraderie of my excellent shipmates. I suddenly have the urge to dive in and do so without thinking. I emerge with a joyous shriek, realising that I too, now, feel at home in the river.

Chapter 12

I roll out of the bell tent, pee behind the bushes, take a large swig of water, grab my tracking pack from the laundry, and wander directly downstream towards the wetlands. Everything is dripping with the last of the wet weekend weather, spider webs hung with silvery droplets from last night's storm. Two trails on the lawn have been made since the dew fell an hour or so ago. I wonder if it could be the two magpies watching me from over yonder? I stop and watch their lanky waddle. Yep, same straddle width. Best guess: fifteen minutes ago. Around the back of the raspberry bushes, I find a handful of dark ripe fruit and take it to the water's edge to munch on. The wetland is also foraging breakfast. A cormorant appears near me with a splash and splutter and scoots into the middle. It must have been holding its breath for some time hoping I'd move on, and so I do, wandering down the path until I find a good muddy spot.

Unlacing my left boot, I let my foot hover a few centimetres above the ground before landing two deliberate prints. The mud is sticky, and clings to my foot with cool tenacity. Circling back, I squat on a log and peer down. My footprint feels both very familiar and yet strangely foreign. The smallest three toes look like pea pods snuggled up neatly next to each other, the big toe almost graceful in its wide statement. The arch barely touches the earth, the heel a perfect U-shape. The second print has an enlarged crest of mud on the outside of the last two toes, evidence of the small twist to the right that I made pre-empting my return. As instructed, after some minutes, I pick up my pencil and notebook and try to sketch the track from memory — the lines, the shapes, the ridges and crests. It's amazing how quickly the detail that moments ago seemed unforgettable has become hazy. I cheat, glancing down before I can stop myself in order to see again the line running under the ball. That's right. I'm pencilling in the blanks when a large black shape suddenly sidles up next to me in the form of a shaggy wet dog, which is quickly followed by a matching creamy one. The yin and yang of the canine world are treading all over my tracks. The owner appears around the corner, an older guy with glasses, blue sports shorts, and ankles flecked with mud.

'Kirt!' the owner says with gentle admonishment to creamy as she snuffles all over my sketchpad.

'It's okay,' I say, hiding my annoyance.

'Well, this is an unusual sight,' he says, pointing down to my single bare foot and sketchpad.

I consider not explaining myself, but I'm feeling somewhat generous to the world after an hour of early-morning wandering.

'I'm learning tracking,' I say simply, expecting him to nod politely and walk off, but instead his interest grows.

'Ha, wow, that's an unusual hobby,' he smiles. 'Are you wanting to track yourself?'

'Well, kinda. I'm practising a few things here, including track ageing,' I say for ease of simplicity. I'm not about to launch into the details of the exercise, which includes strengthening the power of recall.

'How can you tell the age of a track?' he asks, leaning down to look at my footprint. Only a master tracker could read my soul through my sole, but it still feels a bit invasive.

'I guess sketching my foot in different substrates is part of that,' I say. He nods. 'And then I'll come back later in the day and tomorrow to see how the track has changed.' The when question is one of six 'elders' of tracking that I'll focus on in turn, the others being who, what, why, where, and how.

The guy shakes his head. 'It surely must be one of the lost arts.'

'I guess it is.' I smile back. Tracking is an art as much as it is a science, and in its advanced form is pretty much extinct except in a few tiny pockets around the world. I began apprenticing to tracking more than ten years ago, when I studied with Tom Brown Jr, one of the world's most famous contemporary trackers, himself trained over a decade-long-period in the 1960s by Southern Lipan Apache elder Stalking Wolf, one of the last remaining master trackers and shamans, who helped his small people avoid reservation life by hiding in remote mountains. My apprenticeship was shared by the other hundred keen students who gathered in the sandy pine barrens of New Jersey, all of us willingly putting our noses to the ground for hours on end in order to try and even get a glimpse of what seemed to us like invisible evidence of animal movement across the landscape, but which to Tom was like red flags waving in his face. The more I observed of him, the more I realised his encounter with the ground was entirely different to ours. What he saw was not just substrate — grass or leaves or sand or mud — but an animated and ever-changing map criss-crossed with paw prints and hoof marks and claws and scrapes and raccoon skirmishes and families of deer feeding on wintergreen and mice running to evade the cat stalking in the bushes not far behind. He saw the earth as story. My attempts to read even a small part of this story felt both as difficult and as imperative as learning an almost-forgotten language, one older than words.

Finding applicability has never really been much of a motivating factor with me and tracking. Even less so now I live in the city. What use is it to be able to read the age of the dog prints in mud by the river? Still, this pull to return to the dirt feels like it has something for me that doesn't really have much to do with sketches of my feet.

My tracking renaissance is in part because I've engaged a mentor — Josh Lane, a tracker from upstate New York who was mentored by Jon Young (who put me in touch with him). Rather than thinking about tracking as the act of studying mammal prints in mud, Josh is redefining it as being super curious and actively investigative about nature in general, inviting me to use all my sensory capacities to solve mysteries and make connections across the landscape.

'Holistic tracking,' he says, 'is the virtuoso art of the naturalist.'

It's not so far from what I've been doing, just amped up a few notches. The mentorship is a three-month stint involving daily practices and homework as well as fortnightly Zoom calls. I even have a workbook to keep me 'on track'.

The first call with Josh put a bomb of curiosity under me. After telling him a bit about what I was observing, he looked at the photos of my sit spot that I sent him and the questioning began.

'Have you mapped and named every plant within one hundred metres of your sit spot?

'Which direction does the cold wet weather typically come from?

'Do powerful owls keep the same partner each year? Do they always roost in a tree hollow or just when nesting? Are they calling this time of year? Do they feed on the flying foxes? If so, what time of day or night would they do that? What are the flying foxes eating?

'What time of the year do the foxes den in your area?'

I ummed and ahhhed and pondered and wondered.

He encouraged me to start documenting scat. I sent him a photo of a splatty plum-coloured poo in the middle of the path not far from the back door. I was sure it was cat. What else could it be?

'Does a cat eat fruit? Can you rule out possum entirely? What about fox?' The questions began to burn holes in my surety. It's easy to be the best tracker when you only ever track alone.

'If you didn't cotton on already,' Josh said, 'asking what a creature eats is what we like to call a sacred question — one that leads to many more questions.'

At Josh's suggestion, I tried out different 'triggers' — semi-regular events that would be my reminder to stop, tune into my senses, and listen to what was happening both inside and outside of me. I tried his suggestions — the sound of a plane going overhead or the act of walking through a door — but in the end, I came up with my own: whenever I heard the call of the grey shrike-thrush.

I came away from the first call electrified with inquisitiveness. I wanted, no *needed*, to know the answers to my nature mysteries, not in order to get a gold star from teacher, but because there was someone to catch the stories. My sit spot has never known such attention — I'm down there daily, without any effort. As well as regularly springing out of bed to find the best mud puddle downriver, as this dog walker can attest.

I suddenly realise how cold my one exposed foot is and start dusting it off with my sock. The guy is wrangling blacky and creamy but continues the conversation.

'My grandfather was an amateur naturalist,' he says. 'He died of cancer when he was eighty and attributed his cancer to studying blue-green algae under the microscope.'

'It's not free diving, but being a nature geek has its dangers, like too much time alone,' I joke, and we both laugh, but it's not far from the truth. I've not yet seen anyone else at 6.30 am on a Monday morning eating foraged berries with one foot in the mud and a pencil behind their ear.

'I try to remind myself to put my head up when I'm walking,' he says, 'because I often find I'm looking down.' He laughs at himself.

'And looking down means in your thoughts rather than being where you are?'

'Correct. Do you find your thoughts slow down when you're doing this?'

'Yep, I do.' But I don't want to explain the feeling of flow I sometimes get when I'm absorbed in a nature mystery, when all my senses are engaged in a place and time. It's like a kind of ecstatic mindfulness but effortless, with very few thoughts to contend with.

'Well, at least I have a great rock collection,' he says. 'They just sit around and gather dust. But I like finding them.'

'As long as you're curious about something …'

We pause, acknowledging the connection made.

'Why tracking?' he asks, and I'm caught off guard, thinking we were winding up the conversation. It's a question I've been asking myself too. Why am I picking this back up again now, ten years on, in the city? What's the point?

I pause, looking at the guy as my thoughts take me back to something Tom Brown Jr used to say: 'The first track is the end of a string. At the far end, a being is moving, a mystery, dropping a hint about itself every so many feet, telling you more about itself until you can almost see it, even before you come to it. The mystery reveals itself slowly, track by track, giving its genealogy early to coax you in.'

The time my nose was closest to the dirt, when day after day I studied tracking with Tom, was also the time I felt closest to finding some essential part of myself. It was a time when shadowing a trail of fox prints in leaf litter seemed somehow related to the trail of synchronicities I was also following. Life had a momentum to it. An intrigue. Like it had been sprinkled with magic dust. It was an era when the unexpected became expected, the numinous almost normal. The flow state I experienced had something to do with the simultaneous attention to excruciating detail and acres of wild country, something to do with the way tracking trains you to notice the invisible as much as the visible, the intuitive over the rational, and something to do with the lineage it has been passed down from. Another of Tom's infamous sayings floats back to me: 'The place where you lose the trail is not necessarily where the trail ends.'

The night after my first call with Josh, I had a dream. My friend had taken an Aboriginal baby girl away from her home, and we were looking after her. The girl asked when she could go back to her mother. Being out here again with my footprint in the mud, I'm picking the string from that time back up again, wanting once more to find that mysterious being of myself at the other end. It's definitely not the answer the dog walker is looking for. Instead, I offer him a certain slice of the truth.

'Tracking allows me to peel back the layers of the landscape a bit more, learn about the patterns of nature in my neighbourhood.' He tips his head to one side, contemplating. 'You know, see how everything fits together.'

'Hmmm,' he says almost wistfully. 'Lucky you.' And squelches away downstream with yin and yang in tow.

The morning after another call with Josh, I approach my sit spot with a renewed slowness, as if Josh is watching me with his words in my ear. 'Walk five or ten steps, and then stop. Look left and look right. Look up and over your shoulder. Listen to the sounds around you. What is making the noise? Look high in the sky. What are the clouds doing, and what are they telling you about the weather that has passed and the weather to come? Look deep into the shadows — even at midday — for signs of animals and movement.'

All year, I've been aware of a narrow trail that passes through the middle of the tree violet on one side of the path to the river, and then through a hole in the neighbour's fence. I get down on my knees and crawl towards the fence to where the trail disappears into the blackberry thicket. I lower to my belly and peer in. It's almost completely devoid of sunlight, dank and damp.

'Remember Ingwe's words,' I recall Josh saying, quoting a famous British tracker from South Africa: 'To walk in your own footsteps means death; the leopards will wait for you.' The leopards of Fairfield have given me a wide berth so far, but breaking up my routine has definitely been bearing fruit. I've been stumbling upon clues to two of the ambitions I set out on the mentorship with. The first is finding where the foxes den. The second is finding where the powerful owls roost. The largest of Australia's forest owls, the powerful owl is a hunter of the highest order. Despite its size, this is a creature not oft sighted; expertly camouflaged and with numbers waning to endangerment, its presence is usually only known, during certain seasons, by its bassline call — the one that makes the rest of the forest freeze in goosebump terror.

'How big a den am I looking for?' I asked Josh on the call, thinking back to my childhood, where the fox dens in the paddock were conspicuous multi-entrance caverns.

'As big as their whiskers,' Josh said. 'As a general rule for any animal and their burrow.'

Josh had me 'air-sculpt' the shape of a fox while I pictured it in my mind's eye. It felt a bit silly, but I tried it anyway and was surprised to find the invisible fox to be much smaller than I had conceptualised, its shoulder girdle as slim as a large cat's.

'And do they even bother digging a den in the city?' I asked him, telling him about how I had fielded calls on ABC Radio Melbourne last week; one woman had rung up saying a fox was denning under her house.

'How many generations does it take to shift a behaviour? That's a good question,' he said in reply. It was his way of throwing me an edge question — the answer just out of my reach without a good deal more research and observation.

Just as I'm about to pop out into the riverbank clearing near my sit spot, I glance over to the fence line. I see the ears first, two perky little doe-like ears sticking up above clump of poa grass that the animal's attempting to hide behind. It freezes as do I. We both know there's no hiding and make eye contact. The fox is small and young-looking, his hazel eyes equally frightened as inquisitive. One ear folds back to listen behind him. He holds my gaze for a few more seconds before turning on his tail and slipping away.

What else invisible to my usual gaze would appear if I knew to look for it? The more I learn, the less I'm certain about what is or isn't happening around me. I told Josh how I can feel the fox's quiet, watchful presence around the place, and about our mutual awareness of each other. I'm finding it a comfort to know there's a secret life being lived not far from my own. Like I'm in good company.

'That's what we call tone,' said Josh. 'Tapping into the resonance of the animal we're tracking.'

The fox has been sniffing around my beehive. What business could a fox have with bees? I ask my bee mentor, Emma, to see if she has any clues. She doesn't have any answers for that, but she does give me more tracking homework — to notice what's flowering around the neighbourhood and the colour of the pollen on the legs of the bees as they enter the hive. The most obvious flowering plant is the paperbark on the riverbank. I start sitting close enough to the hive to see the mustard-yellow pollen glued to the legs of the bees as they return from their forage. Emma has some of her own tracking juju going on. One evening last week, we crossed paths at an event. She beckoned to me and pointed into a tree to show me an almost indecipherably camouflaged tawny frogmouth.

'Whatever I'm interested in, I start seeing everywhere,' Emma said. 'It's bee swarms and birds for me at the moment.' There's a sweet spot, it seems, at the lapping edge of focus and curiosity, when what we are tracking starts tracking us too.

I dissect owl pellets (scats) that I find under a tree. Some of them have tiny nail sheaths, and I wonder what they could be. How many pellets do owls cough up a day? How many pellets indicate a common roost site? Questions within questions within questions, like a babushka doll I'm unpacking layer by layer.

I locate a small section of the purest sand about twenty minutes' walk away, almost directly under the major eastern arterial road — useful for self-tracking. One of the practices has me sketching micro-tracks within my prints made from a series of movements — normal stance, relaxed stance, right arm raised, bent over, walking, running, even swallowing. The sprint tracks are interesting: the wave I identified from my normal walk turned into a tidal wave that spilt over the levy banks and flooded my heel with frothy sand balls.

I'm still not too excited about these very geeky elements of tracking, but the experience of intimately hanging out with my footprint is intriguing. More often than not, my sketchbook is sitting idle, and I'm just staring into the track. It's like getting to know myself from an unusual and obscure angle, as if studying my shadow. I love the way my arch almost but never quite completely disappears from view, the thinnest of sandbars between ball and heel, like an archipelago. I can sense the almost infinite number of stories in it. There's the way my big toe curves slightly outwards like my mother's. The narrowness of dad's heel. I can sense the footprint's power, representing the capacity and strength that have carried humans and their evolutionary ancestors across wild country. And it's also the imprint of my essence. Not in anything

I can quantify or even put into words, but no one and nothing else can lay claim to this sole. There's something so familiar and comforting about it and yet unfathomable, like I could never fully know the full scope of this being. The more time I spend with my print, the closer I feel the other end of the string is.

A light breeze picks up the hairs on the back of my neck, the cool autumn tickle tempered by the sun peeking through thinning clouds. The Rewild Friday crew are wrapped up in the ubiquitous Melbourne puffa jackets, in shades of black, green, and apricot. Despite the early morning, they're chattering away like the burbling flock of brown thornbills feeding in the flowering wattles on the side of the trail.

'Hey, let's check out the shelter,' says Laura and holds out her hand towards Elena. The rest of us follow Laura's lead through the young eucalypt plantation to the almost-invisible debris shelter.

'It's holding up well,' says Michelle with a smile. 'But I'm still not going to sleep in it.' It's the first time I've been back since the build. I see some eggshell shards on the ground near the entrance to the shelter but don't say anything.

We're back in Westerfolds Park for some tracking practice. It's perfect terrain for beginners, with mud patches that catch the prints of the big mobs of eastern grey kangaroos hereabouts as clear as plaster. We sneak out to the grassy verge, where a mob bounds past within metres of us. They're clearly used to humans, but perhaps not ones who are intent on studying their micro-pressure releases.

Today's theme is 'trailing' — following a set of compression shapes through different terrain. We started out with the log-drag exercise — teams attempting to follow a trail left by one of their members, who has dragged a roped log through the bush. Sounds easy. It isn't. The corresponding story of watching Tom trail a mouse over wooden floorboards puts the skill in perspective. Tracking is humbling.

'Look at this,' says Laura excitedly, 'you can see the very tiniest detail in these tracks.' I look over her shoulder. It's true: even the hairs on the outer edges of the paws are etched into the fine, silty mud. The crew is almost self-organising these days. Elena and Laura are scouting the ground for starting points. Tas has started measuring and whittling a tracking stick — a straight stick that acts like a measuring rod, with markings on it indicating the average bounding length of the roo they're trailing, which will be invaluable when the mud turns to grass.

The questions have already started. It's music to my ears.

'This one has a stride length of 1.2 metres,' calls out Tas.

'And this one is 1.6 metres,' says Michelle. 'I wonder if it's a male, or just older?'

The three small groups soon gather like flies over their chosen starting track, marking each bound excitedly with paddle-pop sticks. They don't need me now. I wander a bit further afield with my tracking mission. My reason for coming here was partly selfish. Josh gave me my final tracking challenge on our recent call — to find and sketch prints of twenty-five species of mammal.

'Twenty-five!' I exclaimed. 'I don't even know if there are that many in the state!'

'Well, that's a good question to start with,' Josh said. 'Think of it as a tracker's rite of passage.'

'Sounds like a graduation exam.'

'Let's just say you're going to need to get creative.'

Josh sent me a template with categories such as dewpoint, humidity, and wind direction. It's definitely a bit more geeky than is my style, but I'm also excited about the treasure hunt. I started with the low-hanging fruit — cat, dog, and fox. I found the fox prints at the wetlands in the mud, distinct from dog in their more elliptical and elongated shape. An earlier trip to Westerfolds gifted me kangaroo, rabbit, and wombat. That's six. Seven was a bit trickier. I researched bandicoots and found out that while the southern brown bandicoot was almost extinct in Victoria, there was actually a healthy population in a suburban botanical garden. I strategically organised a Rewild Friday excursion to Cranbourne Gardens and spent the day watching bandicoots snuffle their long noses into the loamy soil.

Number eight needed sand. On dusk, I collected a couple of buckets of pure white sand from the stash at the back of the local golf course and filled two seedling trays with it, placing one on the back verandah and one on the garden path, both with a quarter of an apple.

The next morning, I jumped out of bed, eager to see if the bait had been taken. The back-path apple was gone without a trace. I made a note to self: put the apple in the *middle* of the tracking box. The verandah one, however, had been visited by a tiny ballerina!

'Come check this out!' I called out to Min through the kitchen window. She shuffled out in pyjamas with a cup of tea in hand.

'Awwww, look at that! So cool!'

The rear compressions were deeply imprinted in the sand — tiny canine-like webbed paws with four tiny claws. The front paws were like the hands of a miniature pianist, the first two long fingers pointing straight out front and two fingers separating to each side, each finger and metacarpal pad perfectly etched in the sand, with tiny nails and little gaps between the finger joints. I had been expecting ring-tailed or brush-tailed possum, but these were too small. Probably a bush rat or introduced black rat.

'It's like a tiny human,' Min said, looking closer.

'I know, they're really incredible.' I was similarly mesmerised by the delicate intricacy of the familiarly shaped appendages. The front paws really did look like embryonic human hands. Looking a little more closely at them before turning to my sketchbook, I was suddenly awestruck at the evolutionary forces that went into crafting these exquisite little mitts, the refinement over millions of years that shaped the design of the fingers and knuckles: pads and paws to allow life to thrive; to locomote, hunt, and forage; to build or find shelter; to feed young. No other paws but these would do. So specific and yet cut from the same cloth as my own. What separates us is really so small.

It occurred to me that these critters were likely the same rodent species that we trapped last year in our pantry, the same ones I recoiled from. Rather than dirty kitchen thieves, now they felt like kin. Studying the tracks afresh, I imagined the story written in the sand: front paws tentatively hopping up on the side of the box, nose and whiskers to the air, simultaneously sniffing for danger and tantalised by the fruit, before loping over the rim and taking the prize. I felt a rush of empathy for the creature and closed my eyes, sensing into where it might be.

Today, I'm on the hunt for mammal number nine — brush-tailed possum. I hone in on a tree with claw marks on the trunk, and around the back, perfectly inscribed in the mossy mud, is a set of brush-tailed possum tracks, quintessential with its webbed second and third toes (on the back feet) and opposable and clawless first toe (also on the hind feet), which I can imagine have wrapped many times around this thin trunk, propelling the creature upwards. I begin sketching. I'm getting better at not trying to make it scientific or to scale, understanding now that that's not the point.

'Just do a chicken sketch,' Josh told me. 'It's almost completely unimportant what comes out, it's the link being made between impression, memory, and recall. Pattern recognition comes not from a perfectly copied and beautiful track, but from the process of remembering, forgetting, searching, straining, curiosity-building, and starting all over again. We learn more when we become aware of the gaps in our memory rather than from what we actually remember.'

Satisfied with the first sketch, I flip the page and begin the next, this one placing the tracks within a bird's-eye view of the overall area, as much as I know it — marking the gully, the kangaroo grounds, the river, even the shelter. Still, large blanks in knowledge remain. It's the part of the process I've had most resistance to, but Josh sold it to me as a brain workout.

'Your brain has particular neurons that make notes of the places you go. The hippocampus creates memory maps of the routes and paths you take, and special "place" cells make an inner spatial representation of the spots you know well. The somatosensory cortex has dedicated areas that catch and make sense of all the sights, sounds, textures, and other bits of data streaming in from your walks and wanders. These neural pathways get thicker and stronger for those patterns you pay most attention to.'

'Are you saying that if I observe the same river red gum on my walk each day, my brain will create a river-red-gum-tree neural pathway?' I asked.

'Exactly. And if you pause to connect with the tree, the neural web will start to grow and strengthen.' We remember what we place emotional importance or 'salience' on.

The next time I was at my sit spot, I pondered how strong my neural pathways of that place were. I focused my attention on the large river red gum, noticing how the flower buds were in groups of seven or nine. I put my hands on the bark and savoured the sensation, breathed in close to its earthy scent, imagining the olfactory neurons adding to my growing inner map of the tree. I floated up in my imagination and hovered above the tree canopy, placing myself in the wider landscape — streets, river, parkland, industrial complex, bat colony, wetlands, and my favourite oak tree. Layers of image and information populated my mind map until it was completely coloured in. My sit spot felt fuller that morning, plumped out and multidimensional.

Without knowing this parkland so well, my sketch and mental map is not nearly as rich, but still the quick sketch is useful in the same way. I wander back over to check up on the tracking groups. Lucy sees me coming and calls out, 'Are you sure we can't skip ahead?' They've stalled where the mud graduates to grass.

'Stay with it for a while longer,' I say, smiling. The idea is not to skip over a hard track — it's these you learn the most from, even if they're diabolically frustrating.

Elena picks up the tracking stick and shifts it to a slightly different angle. I watch her scanning the grass intently, her eyes growing wider as she double-checks herself.

'I think I've got it,' she finally announces. 'Look here, the grass is compressed.'

'Oh yeah,' says Lucy. 'There it is!' The group crowds around in excitement, and they plant the paddle-pop stick at the impression like a proud flag.

We break for lunch on the grass, which has now dried out in the continuous morning sun.

'I have some news,' Michelle says with a gleam in her eye. 'I can now tell you that the first bird of the morning at my place is a red wattlebird.' The group cheers at this news, remembering back to the first Friday, when Michelle was shocked at how challenging she found the tourist test. 'For a goody-two-shoes, it kicked my butt,' she told me that day. I was a bit worried that it would put her off rather than be encouraging, but Michelle has diligently been at her sit spot and guide books and asking the questions. She's at the stage where the patterns are starting to click.

'And I can identify a handful of other birds from their song and appearance, and some plants. I feel like I'm finally learning the stories of the land I live on.' She pauses, but it's obvious there's something else ticking. 'It's interesting that when you don't know anything around you, it's easy to be lonely, but once you start making those connections and learning those bird names and songs and the plants and the stories of the land, you realise you're not alone. It really means something to me to know that wattlebird.'

Tas nods in agreement.

Elena is inspired by Michelle's sharing. 'Well, as you know, I've moved to a new house and had to find another sit spot, which has been hard. Well, the other day I went back to my old sit spot and was overwhelmed with the feeling of coming home. I burst into tears and hugged the tree. I didn't realise that day-after-day sitting there, I had really been getting attached. Maybe it's the first secure attachment I've ever experienced!' she laughs, with some tears coming to her eyes at the memory.

Lucy stretches her legs into the middle of the circle.

'I have a story from my sit spot — well, actually from my walk spot, because I wander around a fair bit.' A few of the group smile. 'Yes, I know it's taken me a while, but I'm addicted now,' she laughs.

Despite living down the road from the Royal Botanic Gardens, Lucy resisted finding a sit spot for a year, before finally deciding to 'try it out' down by one of the ponds. Afterwards, she told me, 'Oh my god, it's just like Jon Young says. It's more about a state of mind than being in a wild place. I can't believe it's taken me this long.'

'You planted the seed for this story a few months ago, when you brought in that tawny-frogmouth feather that you found,' she says, looking at me. 'It reminded me of the tawny frogmouth that I saw one terribly windy day in the gardens. There were trees coming down everywhere — it was just chaos. I was worried about it, but they closed the gardens the next day because there was so much debris. And then the next time I could get back was a couple of weeks later, and I asked for a sign, and just by the side of the tree was a tawny-frogmouth feather. It was amazing. But the story doesn't end there.'

Laura is weaving some lomandra into string as she listens. Michelle lies belly down, soaking in the sun, with her head propped up on her hand.

'The other night, someone had posted five photos of tawny frogmouths on the Wild by Nature Melbourne page, and I gazed at them longingly. The next day, I went back to my sit spot, and this time found myself drawn to a particular patch of off-limits bushland around the lake. My eyes went up the gum tree, and there it was, in the elbow of the branch.'

'Oh wow,' Tas said, clearly inspired. 'That's amazing.'

'I really feel like the tawny frogmouth led me there!'

'Awesome tracking story, Lucy, thank you,' I say.

'Is it a tracking story?' laughs Lucy. 'Well, I guess it is. I guess that makes me a tracker. Imagine if you'd told me that a year ago!' Her smile is the smile of a dozen sunflowers in full bloom.

The stories take me back to that last conversation with Josh, and his suggestion that the structure of our brains reflects our personal connections with nature in our neighbourhoods, and that over time our neural circuitry will reflect the varied and complex patterns of place. The greater diversity of connections we make to owl and tree and pond and moss and kangaroo tracks, the more embedded we become in the place that we all inhabit, our brains wiring for mutuality. Tracking makes us pervious, absorbent to the field of intelligence all around us. There's an osmosis that occurs. We become more endeared to the place, and I'd like to think the place becomes more endeared to us. Elena's sit-spot tree, Michelle's grassland birds, Lucy's relationship with tawny, my dance with powerful owl and fox, they're all weaving themselves into our mind maps of how we know ourselves and how we know the world. The multitude of patterns in nature reflect the patterns of our own life, magnifying aspects of our own wholeness — the sturdiness of deep-rooted tree, the stillness and hiddenness of tawny, the unapologetic brashness of wattlebird, the mystery of owl. Tracking the patterns of relationship and nuance that make themselves known to us in nature, we too track deeper into the hidden trails within, rewilding the inner landscape.

Everything is what it is by virtue of its relationship with everything else. Opening into greater awareness of the mysteries of the wild other is leading me into deeper relationship with the mysteries of my own life. I am not a scientist in this role as tracker, not an objective observer accruing data. When tracking, I'm more of a story catcher, writing myself into the broader narrative as I piece together the puzzle of who, what, when, where, how, and why. Instead of setting me apart, it weaves me closer. Tracking without imagination is like music without heart. The data, like musical notes, are just building blocks. The true song of the track is made through the overlays of imagination, knowledge, curiosity, empathy, and some texture of innocence.

It's April fool's day. The sun is shining. If I needed an excuse to ditch work and head out on a fool's tracking mission, the calendar has gifted me one. I close the laptop, grab an apple, and scamper down the back trail to the river. I feel like a kid let outside after a week of rain, almost too boisterous in my enthusiasm. Hop from stump to stump! Swing off the limb! Climb the tree! Eat the weeds! Track owls!

Grab paddleboard. Head upstream. Coooo-eeeeee, owls! Where are you? In a hollow? On a branch? Looking at me right now? Give me a clue, please! Pull up, dock, wander. I'm drawn to a tree in a neighbour's yard. Looking down, I see a pile of tightly knitted-together nuggets of grey fur and hair, strung through with white bones. Owl pellets! I pick through the coughed-up matter with little sticks. Three long, skinny bones completely intact. Tiny ribs? Rodent legs? There's a tooth and more bone shards. I can almost hear the crunching of bone in beak. I hold a pellet to my nose: it's musky.

I'm drawn out of my absorption with a sudden image of the neighbour finding a barefoot woman with unbrushed hair, leopard-skin tights, and binoculars sniffing owl furballs on their grass. I giggle out loud at the thought.

These pellets need further investigation. I collect an array of different sizes and wedge them between my feet on the paddleboard. But we're not going home yet. I cross the river, tie up the board, and climb up the exposed roots of a tree. The grass below is long and snaky. I reach the path and pass a walker with earphones who eyes me suspiciously. Failing to give a normal measured greeting, I realise I'm in a slightly altered consciousness, more animal than human, wary, and a bit bemused. I collect two different kinds of possum scat and put them in my treasure container, along with the shell of a brown grub that was half-sticking out of the earth. Are they different possums or just varied ages or sexes? Questions questions, always more questions. I climb out on an overhanging tree, my feet tucked up like a frog. Small black ants crawl over me. Big bubbles rise up from under a log lying in the shallows. The sun kisses the underbelly of my feet and my arms. It's good to be out on this particular limb again — curious, wild, attentive. In this moment, that's all I really want: to offer my wholehearted attention to this place where I live, both the human and the non-human. And maybe that's all the beating hearts want of me too. Perhaps our own deepest longings are also those of the world.

'When you track an animal, you must become the animal,' said a San Bushman elder. 'You feel a tingling ... when the animal is close. When tracking is like dancing, then it is the great dance ... when the springbok heart beats in your ribs, you see through its eyes, you feel its stripe dark on your cheek.'

The grey shrike-thrush has started calling again, and I raise my head and call back. Willie wagtail flaps nearby. Everything, everything has come out to play! I am crazy with love this afternoon. Crazy with curiosity. What rock don't I want to look under? What plant doesn't pique my interest? I feel like I've entered the great dance today. The tracking is synonymous with a wakefulness to all the ways that life moves through me and through everything around me. As I'm falling back in love with tracking, I'm falling back in love with myself. I stand up and balance on the branch over the water. I take one step and then another. Dancing with the ripples on the water. Dancing in the mystery.

Chapter 13

A fairy trail of lights bob their way down the steep switchbacks from street level to the fire circle. Disembodied in the shadows, the lights are actually smartphone torches illuminating the path ahead, their owners joining us by our village fire this Friday evening from the productive fields of their laptops and offices. As the carriers approach, their phones return to pockets, and their faces are gradually lit by the softer light of the flames. I watch the dusk arrivals from my seat on the southern edge of the fire circle, growing ever more amazed by the swelling number of wayfarers. Most faces I recognise from various nodes of my social web, but some are unfamiliar. They look a little disorientated, as if they have unwittingly caught the train to Hogwarts.

It's Samhain — an ancient Celtic festival marking the end of the harvest season and the beginning of winter, halfway between solstice and equinox. It's a time known as 'the day of the dead', when the doorway between the 'otherworld' opens, allowing supernatural beings and souls of the dead to come into our world. Traditional rituals surrounding Samhain include bonfires, dancing, feasting, and building altars to honour deceased ancestors. Christian missionaries co-opted the occasion and turned it into Halloween, which has all but lost its spiritual gravitas through commercialisation.

A few months ago, I proposed to my housemates that we create our own Samhain ceremony with a twenty-four-hour sacred fire. The intention being to provide a community space to honour ancestors and those who have died and, more broadly, to allow the expression of grief in all its forms. Ceremony is a foreign-enough concept for our culture. Public displays of grief even more so. The Facebook invite is public. Friends of friends are welcome. I was nervous when editing the words, wanting it to be clear that this will be an experiment. 'Grief and sharing sacred space can look many different ways — singing, storytelling, sharing, crying, laughing, dancing. This is an invitation to be with your own and each other's grief, personal and shared.'

We've been here since dawn, welcoming a slow trickle of visitors to share the fire circle. It's been pretty quiet all day, but I have a feeling that's about to change.

I watch one woman walk slowly down the path in nondescript dark jeans and a black jumper. She looks vaguely familiar. Even in the half-darkness, her face appears heavy and her shoulders slumped. In one hand, she holds a bushel of gum leaves. She stands still and takes in the scene. Before her, a blazing fire is ringed by fifty or more cross-legged figures sitting in loose concentric circles facing the centre, some silent, some quietly chatting. The fire circle is demarcated by a ring of branches and the mini rock wall to the north, as if we're contained within the ground nest of a giant bird. In the west, a collection of stumps sit shoulder to shoulder, adorned with candles, photos, feathers, and other small objects. A weatherworn branch stuck into the ground behind the stumps flutters with small pieces of paper inscribed with intentions and prayers. The woman watches as another recent arrival moves inside the circle and positions a black-and-white photo on one of the stumps. I crane my neck to see the grainy image of what looks to be her grandmother dressed in early-last-century formal English attire. The sounds of wood splitting and swamp hens honking echo through the river valley.

Min stands at the eastern entrance, wearing a long, flowing floral skirt and holding an enormous wing in her right hand, its feathers mottled browns and darker shades of grey. In her role as the waterkeeper, the one who welcomes all to the fire, she beckons the woman to come over. The woman approaches Min slowly, tucking her bob behind one ear. Min smiles encouragingly and folds the wing under her arm as she makes several attempts to light up some sage leaves in a bowl. The wing is so large it almost brushes the ground, and my gaze follows it protectively. It was a gift from my friend Elizabeth for my fortieth birthday, taken from a roadkill wedge-tailed eagle near Uluru. Once the sage is alight, Min resumes her graceful pose. The wing almost entirely covers the standing form of the woman, who jumps ever so slightly when the longest feathers brush her arm. Min takes her time to brush the sage smoke over the woman and then holds out a small ceramic urn of water, indicating its use. The woman gingerly dips her fingers in and brings them to her forehead. A small drip rolls down her nose, and she quickly wipes it away. Min leans forward and whispers in her ear. The woman nods, takes a red bandana from the pile, and ties it to her arm. It's the symbol of silence, and this is only the second time I've seen someone in the last twelve hours adopt it. Min looks over at me and smiles with a wink before turning back to the woman and whispering directions towards the bowl of chopped gum leaves. The woman takes a pinch and enters the fire circle, slowly weaving her way through the seated figures. Some people draw their knees and feet up to make space for her and nod in acknowledgement at the red armband. I watch as she holds out her hand towards the fire, her fingers gripping the leaves as if not wanting to let go. Eventually, she releases her grip, wincing as she does so. Is it the heat? I'm not sure. The leaves make a small skirmish of fireworks, which die quickly. With that last formality complete, the woman looks around furtively for the shortest route to the back of the circle, where she takes a seat on a rug, joining the shadows of the coming night.

There's a shift change inside the fire circle. Serra stands up and starts stacking wood in the fire. It crackles hungrily, and the front row shrinks back. Min now has a queue, and the circle expands to a fourth ring. The flames rise up to light those in the back row. Cat enters the circle, her long red hair stark against her black dress. She hunts me out and without saying anything snuggles in beside me. I put my arm around her protectively. Beth is sitting at the back of the circle, wearing a shroud of black netting. I catch her eye and nothing else needs to be said. Her mother is dying of cancer and her eyes are full of grief. The light shifts again, dusk emptying out its last splash of colour to make way for the coffee black of night. The low murmurs slowly subside without prompting. Chelle is sitting on the other side of me and gestures for a guitar to be passed around the circle. She strums a single chord, looks over at me, and smiles before nodding as we both begin to sing a Rumi poem, welcoming all to this dusk ceremony.

Come, come, whoever you are, even though you've broken your vows a thousand times, come, come again. Others know it too and join us, and like a slow wave the whole circle becomes one voice. Chelle speaks the second line, the devotional Sufi chant, and the volume increases. *Yah Mevlana Ru Allah, La A la Halil Yalah.* Night deepens as we sing, welcoming in the last stragglers to the fire, weaving us together in one harmony. Chelle brings the song to a slow close, resting in some moments of silence before opening up the circle to whoever would like to share.

An older man I know as Mike sits up. His long grey beard lights up against the dark of the shadows behind him.

'I'm struggling to know what to do with my fear about climate change. I'm asking myself what is appropriate in the face of that knowledge, and all that I come back to is love. Love is all I can do, and so I sit here in love.'

We hear you, the crowd offers back as Mike settles into his seat. His partner, Sarah, sitting beside him, is inspired to move forward next.

'I don't know what it means to be connected anymore. I once felt a circle of friends sharing life with me, and now I look around and there's no one. We're all so disconnected. There's a wail inside me that can't find a way out. I'm going to try and find a place for it here.' *We hear you.*

I look over to where Laura and Elena are sitting together. Laura is grieving the death of her beloved dog Apache. She looks up at me with tear-stained eyes, and I meet her gaze with compassion. It's good to see them here.

Rakaia, a friend from the country who has travelled in to attend the ceremony, stands up, her dark hair matching the depth in her eyes. There's an intensity to her tonight, which amplifies as she speaks. 'I've been scared of my power all my life. Of my connection to the earth. I've made choices from fear. My great grandmother from the Americas was full-blooded indigenous, and I've been too scared to claim this as my own heritage. And now I'm ready to step into her shoes. To love the land and my children like she did. To be that wild woman.' A rousing cheer ripples through the crowd as she takes her seat.

Nathaniel is obviously moved by Rakaia's story, and he clears his throat and sits up on his knees. A new friend, Nathanial took a break from the ceremony with me this afternoon to paddle up the river, where he told me a little of his work as a mediator in an Aboriginal community in the desert. I know he has returned burnt out, but he hasn't spoken to me of grief. He brushes his blond hair back from his forehead and looks across the fire, appearing almost otherworldly.

'My Aboriginal mother just lost her son to suicide. For three weeks, she lived on sorry-business land, grieving, sharing stories. There were always people there with her. Sometimes you could hear the wailing from a mile away. Sometimes cackling laughter.' He smiles. 'As part of the culture of grieving, my mother had to give away all her belongings. Everything. It marks an ending and a beginning. When I said my last goodbye, I gave her a blanket and tobacco, her only requests. Part of my heart is still out there in the desert, but I can't go back. I grieve for them and for myself.'

I'm struck by the wisdom in the grieving ritual, the stripping of all belongings, a symbol of how grief strips us of all that is and was. It's also a technology of connection, the griever now reliant on their community to provide for them for a period of time. It's a fierce renunciation that makes physical and visible the vulnerability of grief and makes sure that the griever remains connected. There is no possibility for hiding. In this way, grief is a way of strengthening community interdependence. It's an approach that's almost incomparable to the desperate solitude of grief that I and others have experienced following a close one's passing. Some funerals I've attended purport to have it all stitched up at the end of the wake, the mourners sent home without recourse to further community gatherings. Grieving becomes a private pathology.

The inspiration for the sacred fire arose from a ceremony that I was part of in the US where a similar fire was kept burning for four days and nights as part of a large gathering. It was based on the traditional Haudenosaunee grieving ceremony whereby a fire burns for four days and nights following a death in the community and provides a gathering place for drop-in mourners to share tears and stories, food and support. The fire at the gathering was always well attended, and the group swelled in numbers at dawn and dusk when songs were sung. What was even more interesting to me than the ceremony itself was that even when I wasn't in the circle, the knowledge that it was there and that some portion of the village would be sitting in reverence anchored me. It was a sense that permeated the entire workshop. It's as if we were held by the hand of an elder, the elder of an age-old ceremonial practice. There was somewhere to go to be quiet and reflective that wasn't alone in my tent or behind a tree, a place that was prayer embodied without it being overt or overbearing. Perhaps churches provide the same function for some, a place to be in shared non-ordinary time. If the festival village was a human body, the sacred fire was the navel. The sacred-fire experience taught me how critically important it is for cultural repair to reinvent such shared ceremonial containers. 'They're as old as time,' one white-haired woman told me as I sat next to her on the woven rug. 'We barely limp along without these places to connect with spirit.'

The sacred fire holds this intention for four days, sorry business for three weeks. Death in these contexts is an opportunity for greater intimacy and reciprocity, for shedding and transformation. Grief is not something to 'get rid of' — the quicker the better so as to return to being a productive and functional member of society. Instead of a nuisance, it's actually revered for the connective tissue it provides a community. Rather than a private affair, grief is a collective experience, a shared responsibility, held with equal import to celebration rites, its inevitability and universality woven into the cultural fabric of a healthy culture.

Jon Young's words come back to me. 'When someone dies in an unhealthy society, it leads to isolation and mistrust. When someone dies in a healthy culture, it brings people closer together. Think of grief like this: if a forest's fuel load isn't reduced on a consistent and regular basis, then severe fires eventually burn throughout the land. Alternately, humans can reduce the fuel load on the landscape by lighting smaller fires at the right time of year for the right reasons. These kinds of fires help regenerate the land.'

Fires like these. In backyards. In a room lit by a candle. In contemporary ceremony grounds where we gather to regenerate culture, and by doing so, regenerate the land.

We need these rituals more than ever now. Because there is a grief specific to this time that is a great unspoken ocean of tears — that which marks the unravelling of the web of life, the extinction of species, the loss of wild lands, the polluting of the waters and the atmosphere. It is, potentially, the greatest collective grief we have ever faced.

A man with short greying hair and a violin in one hand kneels and introduces himself as Michael. 'I listened to you on the radio a few nights ago,' he says, looking at me. 'I cried. You encapsulated much of my desires over more than a decade. I have lived in my ecosystem for nearly twenty years. I got to know my living community: the plants, the trees, the birds, the ocean. I discovered the bounty of food in my backyard: sea spinach, many types of seaweed and fungi. I spent many hours listening to the earth in my place, many hours removing noxious weeds, many hours trying to make my home more available for other beings, like yellow-tailed black cockatoos. I have given up. My beautiful place has been desecrated by "development". The black cockies have given up too. My favourite sites to gather wild food are now under slabs of concrete. I am not a pessimist by heart, but my heart is broken.'

We hear you.

'The plastic island in the Pacific,' chimes in one voice.

'The Amazon Basin,' calls in another.

'The melting polar icecaps,' is offered from the invisible back circle.

Other voices chime in, offering the names of places, landscape features, beings, like a roll call of disappearance.

I'm surprised when my friend Kate comes forward. She's usually shy of the spotlight. Her shaved head, symbolic of the end of her long-term relationship, is covered in a black woollen wrap. With feather earrings brushing her shoulders, knee-length boots, and a long robe, she looks a bit like an owl in human form. An owl who now has tears rolling down her cheeks.

'It's just ... the world ... what's happening in the world. It all just hit me, driving here,' she says, brushing tears away from her face. 'What are we doing?' She looks at me incredulously. 'The world is burning. There was tree after tree cut down on the side of the road and just left — such a waste. It was an avenue of destruction. And I'm digging through my parents' compost, and there's plastic all through it. It's just tragic. What are we doing? What are we doing? And there's no mentors or elders. Where are my elders? Where are they?' she asks, this time with anger mixing with the sadness. 'I don't even know which land I'm living on. I don't know anything about the Indigenous history of this land, the stories. It feels wrong to be claiming it as home. I'm so ignorant.'

Kate's sobs grow, and I lay one hand on her knee. Black rivulets of mascara are rolling down her cheeks. It's a beautiful sight. Tears flow down my cheeks too as she speaks, but they are tears of joy. If I could have wished anything for Kate in the few years we've been friends, it would have been this — to see her heart cracked open by the truth of our times. What an honour to witness this moment, where Kate now gives her greatest loyalty to the entire world rather than a partner or small circle of acquaintances. It's risky loving the world when so much is in peril, but it's also risky not to. It feels like a key has turned in her search for passion and purpose.

Kate's story, like so many others, expresses the historical trauma of colonisation. As well as being recipients of its benefits, white people have been harmed by colonisation too. The concept of 'whiteness' that emerged through colonisation and capitalism has erased and usurped specific European cultures as well as indigenous cultures all over the world. The wound of colonisation is shared, if unequally, and shows up differently, depending on whether you're conqueror, conquered, or a more recent immigrant. Grieving our own alienation from the land and our cultural roots is an important step in healing — acknowledging our own confused feelings and beliefs about what it means for those of us not of Indigenous descent to belong to a place that was stolen by our forebears.

Grief activist Francis Weller, one of the pioneers of contemporary forms of community grief rituals, says there are five contemporary 'gates' of grief, one of which is acknowledging losses for the world. 'Whether we consciously recognise it, the daily diminishment of species, habitats, and cultures is noted in our psyches.' But, he writes, 'Through our ability to acknowledge the layers of loss, we can truly discover our capacity to respond, to protect and restore what has been damaged.'

Joanna Macy names the experiencing of moral pain for the world as one of the most radical acts we can do at this time on the planet, offering us embodied evidence of our interconnectedness with all beings — including humans of all cultures. And from this experiencing also arises our power to act on behalf of our fellow beings. 'The sorrow, grief, and rage you feel is a measure of your humanity and your evolutionary maturity. Like living cells in a larger body, it is natural that we feel the trauma of our world. As your heart breaks open, there will be room for the world to heal.'

Grieving is the first step in reclaiming what we have lost. Mourning the loss of the 'wild other', the deep connections we experienced with plants, animals, and landforms, is what environmental writer Chellis Glendinning calls our 'original trauma', which carries with it all the recognisable symptoms: chronic anxiety, dissociation, distrust, disconnection, depression. The result is a profound loneliness that we rarely acknowledge, or alternatively pathologise as something personally awry.

Human biologist Paul Shepard says, 'The grief and sense of loss, that we often interpret as a failure in our personality, is actually a feeling of emptiness where a beautiful and strange otherness should have been encountered.'

The gate of grief, says Weller, is where we most directly experience the soul of the world, or the *anima mundi*. This is where we recognise that 'the greater part of the soul lies outside the body'. We feel it in our bones that something is amiss. Our larger self is being degraded and eroded. The litany of losses is almost uncontainable, including our own loss of connection with nature.

I remember clearly the moment that grief for the world moved in me. I was at university, and on an exponential learning curve of environmental politics. I was outraged by the statistics and facts, but the reality hadn't yet sunk in. A few of us went out to check out a logging operation. Standing on a stump of a 300-year-old brush-box tree and looking out at the sawdust remains of what was a few weeks ago a hooting and hollering old-growth forest, my knees started shaking, and the whole disaster came into clarity in my body. It was like a shelf inside me collapsed, and the pain of the world tore open my heart. The experience shook me up in all the right ways, a nodal point in my maturity. Grief was a kind of eco-awakening, an embodied experience of my innate membership in the earth community and a recognition that this membership is my primary place of belonging in the world. My centre of gravity shifted.

Kate's bravery in sharing has clearly moved the woman wearing the red armband, who I see has shifted herself into a kneeling position. Her knees are obviously stiff, and she leans one hand on the ground to help herself up. Unpicking the bandana tie, she dusts off her knees, and with gum leaves in hand starts picking her way towards the fire.

Once there, she waits, contemplating the flames. She looks my way, and I suddenly recognise her. She's one of the women who gathered on the day the trees came down in our neighbourhood. She's one of my neighbours. I watch as she plucks a few leaves from the bushel and offers them to the fire. I catch a faint whiff of the same lemony scent that permeated the entire street when the first branches crashed to the bitumen.

There had been a long campaign to save the two mature lemon-scented gums planted on the roundabout at the far end of the street. They were the kind of trees that make a street, their generous shade falling in a wide perimeter, their character even wider. 'Save Us' signs had started appearing around the trees six months ago. A letterbox drop informed me why: multinational company APA, which was also involved in fracking, wanted to assess the gas pipes under the roundabout for corrosion, and the quickest and cheapest way to do that was to pull the trees out.

In the last few weeks, locals kept a twenty-four-hour vigil. I stopped in on my bike one afternoon. The grey-haired, well-dressed woman on shift was seated in a comfortable deck chair next to another older neighbour with a tennis racquet in his hand. A small coffee table was stacked with gifts from passers-by: baklava and biscuits, a tea station and esky for milk. 'Sign up for a shift?' the woman asked me. 'We need some young people — it's getting too cold for us oldies here overnight.'

'Sure,' I answered, inspired to back their neighbourly spirit as much as the trees. My swag was set ready to go, but I came down with a fierce cold and had to pull out. Instead, my only contribution was bearing witness the morning the trees were removed. The group alert came via WhatsApp as I was making my cup of tea: *Chainsaws and mulchers are here, APA plus 40 police.* I cycled down and stood straddling my bike amidst a crowd of about forty locals, who stood behind the stony faces of at least four dozen police lined up shoulder-to-shoulder. When the chainsaws started up, two women next to me started crying and hugging, and some TV cameras moved in for a close-up. To my surprise, a wave of grief rolled over me too, and I fought back tears. It's just two street trees, I told myself, not wanting to cry in front of the cops.

The chainsaw made contact with the first high branch. The branch wavered and then fell, crashing to the asphalt with a loud thump, its bark flying off in all directions to release the pungent lemon scent. At that moment, another woman rushed in to the front line, to stand eye-to-eye with the police, stretching out both her arms in semi-stop signs, her palms above her head and facing the trees. Slight of build, with two small blonde buns above each ear and determinedly clenched lips, she was the perfect target for the TV cameras, which wheeled around hungrily to catch the David and Goliath moment. Another, larger branch came down amidst the roar of the chainsaw, and tears flowed down the woman's cheeks, while her expression remained as resolute as ever.

That is the face of a contemporary warrior, I thought. It's not just two street trees. It's one of the thousands of small cuts adding up to the desecration of all things beautiful and life-affirming. It's another way that profit and greed and expediency is taking precedence over wild nature. The tears that rolled down the woman's face and the strength of the gesture she made with her arms told me she had the heart to bear witness to the earth's suffering, to let in the pain of the world, knowing that it was the very source of her power. Contained within the very act of turning her face towards the destruction was the seed of healing. Right there, at the end of my street on a Monday morning, was one of the most powerfully symbolic acts of the courage to change the world that I had seen. It didn't matter that it was just two street trees. It may as well have been the Tasmanian wilderness or the Amazon jungle. Her long, thin fingers so strongly held in the mudra of defence were simultaneously inside my heart, stretching it wide open.

And now this neighbour, who I have never talked to, has come to say her last goodbyes to the trees, perhaps to grieve the trees that she doesn't see falling, all around the world. She takes a deep breath and places the leaves in the flames. As the thick lemon scent ripples out, she smiles, stumbles up to standing, gives a gesture of prayer to the fire and the circle, and crosses back over the threshold and into the night.

The songs continue into the night, songs of love and loss, devotion, inspiration. Poems too, of despair, mystery, longing. And the stories, oh the stories, poignant as my own: the ingenuity of great-grandfathers in concentration camps, the smell of mugwort and garlic in grandmotherly kitchens, moments of birth, moments of death, goodbyes. Late into the night, a cellist unleashes her bow across the strings like she's singing each our stories, spoken or otherwise.

Suddenly, there's a barely audible flapping of wings, and I look up to see a tawny frogmouth alight the branch directly over the circle. The cello stops, and the circle of fifty people pause to look up. The tawny frogmouth, hazy in the dusk smoke, cocks its head to one side, as if listening. We listen too, in thrumming silence. The cello starts up again low and slow and with it a collective hum, which shifts into beautiful spontaneous harmonies. Who are these wild pilgrims with their stories rich and peculiar as myth? Are they really from the streets of this gritty town?

As I stroke Cat's hair, I am also content as a cat in this soft and strange land we have somehow created for ourselves.

As the small hours of the night approach, the circle dwindles and quietens. Tiredness overwhelms me, and I head up to bed for a nap. My body wakes me at 4.00 am. Time to head back. I pass through the waterkeeper's threshold and enter again the firelight. Half a dozen figures are curled up sleeping; another half-dozen are wrapped in shrouds, meditating or staring into the fire. I lower myself to the ground and sit facing the river. The light is different from two hours ago, even though night still lays full claim to the shadows. I look around as if I'm being watched. The day is padding quietly towards us, I can feel it, waiting behind dark wings. The silence is thinner too — restless, edgy, as though it's listening from the other side of a curtain.

My drum leans up against one of the altar stumps, where I left it this morning. A strong impulse to play comes over me, and despite my reluctance to interrupt the quiet, the urge grows. Finally, I give in, retrieving my drum and taking it to the edge of the fire circle. Still facing the river, my fingers start slowing tapping on the skin and a half-forgotten song I heard years ago appears on my lips.

> I hear the moon and the fox and the eagle, I hear the
>   moon and the fox singing,
> I hear the moon and the fox and the eagle, I hear the
>   moon and the fox singing,
> And in twelve moons, I will return,
> I hear the moon and the fox singing.

The song rolls over itself again and again, my fingers keeping time with the trance-like rhythm until at some point it begins to feel like the song is being sung through me. All effort dissolves, and I'm singing and swaying, bare feet on the earth flowing with water, purring with fire. I am courting the night and it is court-ing me — quick now, before I disappear, it says, dance with me before I fade. The darkness keeps catching and releasing, catching and releasing, the drum now a full moon that I rest my cheek on as the song quietens, but the music continues almost inaudibly, my fingers tapping the beat.

At some point, I open my eyes to notice a crack in the horizon and others seated around the circle. The sleeping figures are stirring. My housemates are smiling at me with love. I'm proud of us all. I don't want day to arrive just yet, though; I want to hang out in this liminal beauty, bathe in the shared pool of all that has been said or sung here in the last twenty-four hours.

A lorikeet calls in the distance, its single screech enough to send the night scurrying out the back door with the stealth of fox. Somebody brings over the branch that's fluttering with hand-written notes tied on with string and lays it on the fire. It's the last timber to go on the fire. The sun crests the trees to the east. With cheek kisses and a loud holler, the ceremony is over.

I breathe a deep sigh of relief. The experiment worked, and with a strength I couldn't have predicted. It's so important, this taking of grief out of the personal and gently working it back into collective ritual. When we attend to the sorrows of our times, we also attend to what we long to reclaim, what we instinctively know is possible. If, as Weller says, we let the losses of the world 'penetrate our insulated hut' of the heart, we are put back in right relationship with the world. For we only grieve what we love.

The roundabout at the end of the street shapeshifts as the weeks pass. The initial flowers and candles on the stumps transform into a contemplative scene of a single chair with a white bedside table, matching lamp, and book. On the road, a Dr Seuss quote is chalked: 'Unless someone like you cares a whole awful lot, nothing is going to get better. It's not.' When the stumps are removed, kangaroo paws and other small native shrubs are planted its place.

The changing landscape mirrors the stages of grief. If you moved to the street now, you wouldn't think it wanted for anything.

Kate returned home determined to learn more about the land she grew up on. She sends me a series of very excited voice messages.

'The totems for the area were the eagle and the raven. Last night I dreamt that I was at home and an eagle landed directly in front of me and dropped feathers. It was like some kind of gift.

'I went into a local art gallery and there was a free lecture about Aboriginal culture and the woman who was speaking was Torres Strait Islander and she's written a book about becoming indigenous to the world and she lives down the road!'

Grief is a doorway to connection. The door opens, eagles and ravens fly in and perch on your shoulder, silver-haired crones put out a welcome lantern at dusk on your street. Poems pour out, songs are harvested from the rivers and mountains. Rituals relate us to each other again and again, reminding us that we are part of a complex web out of which we cannot fall, and to which we are daily indebted the gift of life.

If you have become frozen to grief then you have also become frozen to love. Thaw yourself, speak your deepest yearning to the moon, cry tears into the earth's hollows, be cracked open by the bell-like chime of the first bird in the pre-dawn light. Come, come, whoever you are. Assume your rightful seat at the table of creation. Break bread with the beaked ones, share honey mead with the large cats while snakes spiral up your legs and wrens weave nests in your hair. And in twelve moons, we will return.

Chapter 14

I wake at first light and blink my eyes open. It's too quick for the dream images. They turn on their heels and scurry back down the ladder before I can catch them. I walk upstairs and switch the kettle on. From the crack in the window, the brush of autumn greets my skin. The magpies are warbling. They remind me of my grandmother's garden. Floating clouds are glowing a pale pink. The kettle clicks off. So quick. I fill a pot with leaves and steaming water and continue staring out the window.

Inside or outside. Here's the daily choice again. Do I stay inside and tend to the cultivated life — make toast, sweep the floor, reply to a text message, take a cup of tea up to Min. Or do I instead step outside and give my attention to the outside world, waking up to a very different rhythm? Inside is primarily under my control. Beyond four walls, my schedule flies out the door, and I'm subject to interactions and elements outside my control.

Today is the Sustainable Living Festival, and I'm talking on a panel titled 'Animism Respoken'. I haven't given it much thought yet. I really should make some notes. I stare longingly out the window. The clouds are palms outstretched towards me. It's now or never. Before the plans and responsibilities can wrap their tendrils any tighter around my ankles, I slide the chair back, leave the tea steaming on the bench, and head out the back door and down the hill.

The river is a good few degrees cooler, mist creeping along the surface and gathering like a ruffled skirt at my knees. I lower myself onto the paddleboard and steady myself against the jetty before coming to full standing. With a single swish of the paddle, I'm free of the shore. So effortless is the momentum that I could be floating above water except for the ripples the board makes in its wake. The highway is blessedly mostly still asleep, the occasional roar of an engine like a blast of wind through leaves.

A black-faced cuckoo-shrike flies in, its call a sweet and haunting trill. There is a faraway knowledge of needing to have something coherent to say into a microphone in the city later. My co-panelists, Patrick Jones and Maya Ward, have been ping-ponging ideas over email, and I've felt too busy to put my mind to it. Still, the idea of speaking in front of a crowd feels conceptual compared to the full-bodied experience of propelling myself down a body of water in the early-morning light. Two Pacific black ducks catch up to me and swim alongside like loyal little tugboats. Two ibis spread their leggy wings on a northward trajectory. It's like Noah's Ark, two of everything. If I could, my panel would be given here, with all of us together at river level, watching and listening. I know that's all I want to do this morning.

I paddle myself upstream, my bare feet tingling in the morning cold. I suddenly have an impulse to look up into the elm tree approaching on my left. It's veritably glowing with a kind of iridescent European forest green this land doesn't really know. I wonder that I've barely noticed it before. I swim my vessel a little closer as my curiosity grows. It's a wide, languid tree, planted on the edge of a neighbour's yard. The leaves flutter like prayer flags gently in the breeze. My attention penetrates beyond first impressions to the detail of dark trunk and limb. It's then I become aware of a large dark shape in the branches. Pattern recognition immediately starts flashing me messages — 'Not branch-shaped. Bird-shaped. Investigate.' I paddle in closer, keeping focus-locked on the shape.

A strange high-pitched whistling sound starts emanating from the thickest bushels of branches. As I get closer, I can sense a presence that is definitely not of the vegetable kingdom. The feeling grows stronger as I near the bank and steady myself against an overhanging branch. When I look up, I'm met by the piercing gaze of two huge yellowing eyes that arise from an enormous bulk of brown mottled feathers. A bolt of adrenaline runs through my body, and I gasp. I know this face. I know these eyes. Standing at least a half-metre high, the owl's talons grip with vice-like strength to the thin branch. It holds my gaze with similar strength, and I recoil slightly at the intensity. It's none other than the powerful owl! The creature I've been tracking for months, the one whose velvety call I've heard night after night, the one whose pellets I have held to my nose and pulled apart by hand. I've found it, or it's found me. Either way. We finally meet.

The owl and I stare into each other's eyes, unblinking. Dark bushy eyebrows furrow into a sharp V as if asking, 'What are you doing here?' Although it doesn't appear to be moving its mouth, the whistling call is definitely coming from the owl. I don't need to speak its language to know it as a warning. Don't get too close. I steady myself against the branch, not wanting to break the gaze.

I suddenly remember that I don't need to remain silent, and if encountering a human it would be rude to do so. I raise one arm above my head in acknowledgement and call out, 'Greeting, owl of the elm — you've finally revealed to me your hiding place!' The owl goes quiet and cocks its head to one side, as if not quite sure what to make of me now I am no longer staring like I'm at the zoo and instead am prepared to enter into a conversation. I extend a heartfelt warmth to accompany my vocal greeting and engage my senses in an all-bodied listening for a response.

The owl cocks its head to the other side. As I continue to stare into the its eyes, its shape begins to change, the face widening and flattening, more human. All of a sudden, it's no longer the owl's face that I'm looking at but a young me. The image is only momentary, but vivid, as if I was looking into my eyes as a four-year-old. A bowl cut frames my round face, large searching green eyes, a small cut on my upper cheek. I look uneasy, a little shy, but fierce. It's an image that still sits in a frame on my dad's mantelpiece. I've always been intrigued by it, the hint of world-weariness, and some ageless wild essence behind her eyes. I've always wondered why out of all the other happy snaps, my father chose this one.

Slowly, the image fades back to the fierce feathered face, the eyes still unblinking in their gaze. Maybe the owl is encountering me in the same way as the young girl did the photographer — wary and self-possessed. The initial ecstasy of connection with the owl fades into a dense kind of shared silence infused with the intensity of my small child's gaze. I feel like I've climbed into a tree hollow and entered a different reality. The energy between us amps us. I hold my breath, knowing something is about to happen.

Without breaking my gaze, all of a sudden, the creature begins making gestures with its neck, a horizontal back and forth, back and forth, again and again. What is this dance? I am intimidated and seduced. It pauses for a second and then resumes, left and right, left and right, in rhythmic time. It's drawing me in. I can't remain static. I begin to mirror its movement, moving my neck from shoulder to shoulder, keeping my chin parallel to the river and my eyes interlocked with its. We move as if a mirror image of each other, back and forth, back and forth. A hint of challenge passes between us. Who dares to break this tryst? The owl's eyes seem to narrow and invite me in deeper. I begin to enter a light trance, the physical features around the owl blurring. We move together, and all else fades around me, while the owl comes into clearer focus. The sun rises and makes contact with the east-facing elm leaves, and the sudden contrasting light makes the owl appear even darker, as if its fading into the background along with the night. The shadows bring me deeper into the owl's gaze. We remain locked on each other, as if besotted, engaged in some kind of courtship rite. And just as quickly as it started, the dance slows to a pause and then a standstill.

From close by comes the sound of flapping wings, and my gaze is broken by the appearance of another of its kind, flying in between us and landing on an adjacent branch. My eyes track this smaller shape to its perch. A juvenile! And then another smaller owl flies in. Three! Three powerful owls! My legs are shaking ever so slightly, and I remember I'm floating on water. I call out a greeting to the two younger owls, feeling sure that I've passed some test.

The parent owl starts ruffling its feathers and looking around, restless. I don't want this to end, but recognise when it's time to take my leave. I put one hand over my heart and speak some words of gratitude. The three owls stare down at me motionlessly, as if saluting me on my way. With a wave, I push the board from the bank and float out into the centre of the river. I look back, and the three are already almost undetectable — dark shapes amidst a tangle of dark branches and a brightening sky.

Riding the train into the city, I can feel my presence emanating with the dark underbelly of the encounter. People are looking at me curiously. They're not sure why, but I know. It's not mine, I want to tell them. I don't own it. I'm cloaked in owl feathers for the day. It's shrugged around my shoulders, flecked with gold, and embroidered with thousands of moth wings.

I try to bend my mind towards what I might offer on the panel discussion. I turn to Google on my phone. 'Animism is the belief that objects, places, and creatures all possess a distinct spiritual essence, that all things — animals, plants, rocks, rivers, weather systems — are animated and alive.'

To what extent did my belief system allow for the owl encounter this morning? Would I have been that open and responsive to a rock or a tree? What are the limitations that I have self-imposed? And where is the line between conversation and fantasy?

In earth-based cultures, the kind of interaction I had with the owl would not be considered unusual, not a 'religious' or special experience but an embodied extension of the philosophy that a community of relationship exists with all organic life. Everyday conversations with wildish others were punctuated with less frequent encounters that held greater significance — experiences of the numinous, like the shimmering moment I shared with the owl this morning.

Numinous. It's a beautiful word, but not one in common vernacular, its rarity reflecting the margins by which the experience sits within our cultural frameworks. It's a word that derives from the Latin 'numen', meaning 'a nod'. It's a sweet etymology, the idea of 'a nod from god' — an address from the incomprehensible. A numinous encounter might contain qualities of sublimity, awe, rapture, and allurement, but also potentially fear, repulsion, bewilderment, or humility. Whether it be a dream, waking vision, or significant encounter with a non-human entity, a numinous experience is recognised by its luminosity, its extraordinariness, the meaning or wisdom it may offer one's life. My experience this morning had an element of the numinous in its otherworldliness, and in the feeling that we had been swimming in each other's psyches for some months now. What could it mean for me to have seen my four-year-old face? Although 'found', the owl continues to be a mystery.

I fish a pen out of my bag and start jotting some notes about this morning's encounter. To my surprise, I find the language of romance emerging: 'flirtation ... passion ... desire'. It's true, the whole interaction had an air of wild eros to it — not of a sexual kind but an embodied ecstasy, a shared electricity at both being seen and seeing another in a moment of unmasked truth, a resonance in both the polarity of our difference and the mutuality of our common heart, a yearning to move towards greater intimacy. The falling-in-love vitality has been palpable all morning, infusing even everyday inert objects with new interest: the toast on my plate holding some deeper fascination, the kettle boiling affably.

Why would I offer eros monogamously to the human world? I feel capable of falling in love with the myriad of life forms all around me, flirting equally with silky moss and storm sky and milky sap and spongy fungi and tumultuous ocean. Possibilities for engagement are endless when I open not just to the linguistics of human form but the multi-conversant language of life, a courtship with the very animate force that pulses in and through all things. It's a tantric flip in perspective.

I step out of the train into the busy station. As I navigate the crowd heading down to the Yarra for the festival, I'm aware of how expanded my presence feels. I greet Patrick and Maya with a hug, and we pull some milk crates together up the back for a quick planning chat before taking to the stage. Behind me, I can feel the density of the office towers and hear the combined hum of trams, trains, and traffic. The gaze of a hundred or so people sitting in plastic white chairs stares up at me. The crowd swells as the sound checks are completed. I look out over their heads to the brown ribbon of river, the same one where upstream some hours ago I was balanced barefoot on a board conversing with animate life in the elm.

I'm relieved when Patrick takes the microphone first. His bag sits open next to him, and I can see his half-eaten sandwich wrapped in beeswax, the homemade sourdough bread and pickles familiar to me from our previous shared meals. It would have taken him almost three hours to get here on public transport from his village home. Patrick crosses his black-jeaned legs and clears his throat. As usual, he is casually but intimidatingly eloquent.

'My understanding of more-than-human consciousness and the consciousness of animals began as a child growing up in a house of dogs. Dogs were my way back into what I call the living of the world. Dogs and humans share a common cosmology. In many Aboriginal cosmologies, dog and human dreamings are the same thing. We see this in other cultures with jaguar and werewolf. And this person–wolf, person–dog, person–dingo relationship is very old and very sacred. I didn't have to learn dog culture. As a child growing up, the kinship relationships were explicit and not academic. My grandmother's advice was to let dogs lick your cuts, and now we know of the incredible healing properties of dog tongue microbes. It wasn't until I moved in with a girlfriend who had a rottweiler named Gustaph that I learnt that relationships between dogs and humans have to be negotiated.'

The crowd is still and attentive.

'I was home alone with Gustaph one day in our little shack, and all of a sudden he came over, stood over me, puffed out his chest out, saying where do you fit in? His presence and muscularity caused me to do something I'm not proud of. I slowly got up, moved above him, screamed in horror and fear, and kicked him out the door. That was my way of asserting my dominance. I hadn't learnt the necessity of working out our relationship with each other. That was my first big learning.'

Patrick looks down and opens to a tagged page in a small moleskin notebook.

'I wrote some notes on the bus on the way here,' he says somewhat shyly. 'Is it time to return to the living of things? To attend to our domestication and sense again what we once understood about more than human life and semiotics? The speakings of wild apple trees, the songs of cicadas and creeks, the signs and manifestation of slime mould and the espionage of mycelium? This is our common sense, nearly entirely lost to religious dogmas, selfish genes, and other Darwinian ideologies, including capital. In the re-animating of the worlds of the world, we will again inhabit calendula time, be intoxicated by gum leaf, pushed around by promiscuous wings, caught by a thick flood of wrens, and licked by the giving and getting love of dog.'

The poetry of the words has lit a fire in my imagination. Patrick is expressing the same eros I've been thinking about all morning. And he's right: the highly domesticated sphere of pets offers us a way in, an accessible opportunity to reperform relationships with not just animals but fungi and plants and rocks and mountains.

In the minute of audience applause as Patrick closes his book, I remember an experience my friend Sam shared with me earlier this week, an experience that similarly rocked her world. She had taken herself to sit under a river red gum and was crying into her hands with the ache of a heartbreaking decision. A kookaburra flew in to perch on the branch above her. Sam's tears paused as she gazed up at the kookaburra and watched on in amazement as the kookaburra turned its head, bit a feather out of one wing, and dropped it directly in front of her. 'It literally bit the feather off! I can't believe it!' she told me. 'What a gift of trust!'

Suddenly, the microphone is in front of me, and I bring myself back to the faces looking expectantly up at me. For a moment, my mind is blank, but then, looking out at the river, I know the story I want to tell. I speak as if it's a story, a myth, envisioning myself there, hoping I'm bringing people along with me as I look up into those yellowing eyes. They seem to be, some with their hands holding up their chins, some leaning forward in their seat.

'It's my understanding that the place that I experienced this encounter was from the deep imagination. This doesn't mean fantasy. It's the kind of consciousness that is able to tap into the magic that lies beyond the veil of ordinary ways of perceiving and understanding. It's the translator, if you like, between the knowable and the unknowable world.'

Sharing Sam's story of the kookaburra feather, I say, 'The feather represented for Sam the gift of trust, a reminder that we're woven into a mystery beyond our comprehension and that there are allies and clues available to us if we choose to open to the greater field of intelligence all around us. Familiarity with this capacity would have been part of a nature-based childhood, but growing up in human-dominated landscapes and psychological landscapes of the rational and inanimate, it's very much a latent superpower.'

I was worried that telling the story of the experience would feel like a psychological dissection, but instead it feels good. Engaging storyteller's mind involved being in my imagination as I recalled the luminosity of the moment as well as in my rational mind to string it all together in an eloquent delivery. Sitting up here on stage, I realise that weaving the two is one of the tasks and the opportunities of our times. We need to be able to pivot between ways of knowing and being in these times, to leap nimbly across chasms of the timeless and ancient, imaginative and deductive, nature and culture, twining together the gifts of the scientific rational legacy and the wild imagination. What synergies might exist in this nexus? It's a landscape still dominated by a cultural addiction to linearity but ripe with possibility.

Maya takes over the microphone. 'The culture that has overtaken the world is a certain way of thinking and has only just emerged. It can't and won't last long,' she says definitively. 'We will return back to a very whole and real way of being. This innate wildness is our birthright, our wholeness. It's true rewilding.'

Arriving home, I have chocolate on my mind and head out on a walk to the shops. My fingers pull against the fine leaves of the paperbark planted on the nature strip, run over the rough bark of the red gums, and graze the tops of grasses. A poodle bounds up to me, and I squat down and rub its neck, much to both dog's and owner's delight.

One woman came up to me after the panel, her cheeks flushed with excitement, and asked me how to 'start'.

'Go for a wander,' I suggested after some moments of thinking, 'Assuming that everything around you is as curious and interested in you as you are in it. And see what might open.'

It's a courtship, I think as I enter the fruit shop and find myself drawn to a ripe custard apple. To let ourselves be wooed by the first songbird of the morning, captivated by the sweetness of dew on tongue, enamoured by the bumpy skin of avocado. Like any romance, there is risk in investing one's heart, in assuming that it matters to another with what words or gestures it is met.

At the base of a street tree on my walk home is a tiny ring of mushrooms, their tops just cresting. I touch the peach filaments of their underskirt, delicate and silky. My heart stirs with affection. Perhaps in this small action I am fulfilling the human species' unique gift of mirroring back to the world its own likeness — the beauty and wonder of life — through the vehicle of love. Simple acts such as these take us back into our primary place of belonging, entering back into the most original of conversations. Yet it flies in the face of the quantifiable, materialist mindset. It's too easy to take, burn, steal, or plunder from something without sentience. It's too easy to override the impulse for care when there is no means of relating. The wild imagination, like the wild organic world, has been colonised, paved over, and stripped.

What if we were to let our imaginations out of the cage? Rewild our psyches so that they were liberated enough to see through the eyes of the peregrine falcon, to climb the grandfather oak and accept the blessings offered within the high branches, to cavort in the mud with the night heron, and be drawn to the tree in an unkempt corner in the neighbourhood that pulses with magnetism? What then? What new stories and alliances might arise? What might crumble?

A few years ago, I dreamt a single image, a banner that stated simply, 'The Imagination Is Dangerous.' The wild imagination is dangerous in the best way. It will not be controlled, tamed, or made to sit on command. It arises from faculties of intelligence beyond our rational thought — not merely from our own wild nature but from the wild nature of the world itself. Dangerous indeed it would be to the primary narrative of our times, which relies on an earth reduced to resources for extraction and wild creatures shrunken to statistics in a triple bottom line.

The modern mind needs to be taken on a short leash. Making plans for Saturday night and updating my list of newsletter subscribers is not going to endear me to the subtleties of fungi poking up through pine needles. It's not likely to induce much response from the maidenhair fern growing in the crevice under the bridge. It's certainly not going to allow much space for the wilder presences to creep up quietly and offer some feeling or image. When I have shrunk the world around me to the role of backdrop for human drama then nothing is going to touch me. I'm unavailable, offline, conversationally mute.

As imagination activist Geneen Haugen says, it would be perverse to think that nature is waiting around to deliver messages about our personal life to us. This too is another way we can 'use' nature. We need first to acknowledge that the wild other has its own longings, its own relationships outside of our sphere.

I think back to something Patrick said on the panel. It was from an anecdote that anthropologist Eduardo Kohn told about his trip with a hunting party in the upper Amazon. He was told by the hunters, 'Sleep face up. Because when the jaguar comes at night and stands over you and looks into your eyes, he won't bother you. But if you sleep face down, he'll think you're prey and attack.'

To be re-enchanted by the world, we need to meet the gaze of the wilder other, with as few projections as possible. This gaze has been a force that has shaped our neurobiology over millions of years of evolution. Meeting it acknowledges that animals represent us as much as we represent animals.

Re-enchanting ourselves is a form of rewilding, challenging the very assumption that we are outside of nature. It will take the faculty of an imagination steeped in charismatic conversation with the world to envision a life-giving future, even one 'beyond our wildest imagination'. Perhaps it is the only thing that can. To fall in love with the world. To be enchanted once again. Could this be the most radical and important act of our times? It starts with the smallest of choices. Like, when the clouds beckon, or the owl calls, or the storm approaches, to say yes, and reclaim our place in the larger conversation.

Chapter 15

'Think of it like unpicking a dress. We're opening it up at the seams,' I tell Serra as we stand, knives in hand, in front of a fox strung up upside down by its ankles to a branch near the clothesline. Both Serra and I have tied our hair up in high ponytails that swing down our backs, similar to the bushy fiery-striped tail of the vixen we're circling. The fox's head suddenly lolls to one side, revealing sharp incisors.

'Ooh!' Serra squeals. 'She's listening.'

'First, we cut around the four ankles and then up the inside of the legs to meet a centre line,' I say. My confident directions are helping to garner my own fortitude for the task ahead. It's been a couple of years since I've skinned an animal, and never a fox. On my mind are the stories my dad told me about how he would sell fox pelts for a few bob as a kid, and how bad they smelt. The pegs on the clothes line might come in handy. I'm steeling myself for the first incision when I hear footsteps out the back path and look up to see neighbour Jeremy appearing, obviously having received my message. He's wearing rainbow shorts and a tank top and comes bearing gifts.

'Protection?' Jeremy says, pulling out an entire box of latex gloves. Jeremy is a welcome addition to the task. Having lived out bush for thirty years, Jeremy once brought an entire cow in for his Steiner class to skin and process. He's handy to have around — in this case, literally.

'Do you want to start on the bum hole?' I ask Serra, feeling a little guilty for palming off my least favourite part of the job, but she seems keen to try anything.

The three of us place our hands on the fur, speaking a few words of gratitude and intention. The fascia is stiff under the soft hair, having contracted in the twenty-four hours since the animal's death. I steel myself for the first incision on the rear ankle, slicing carefully until I feel into the appropriate level of strength, depth, and pressure needed to separate the skin but not enter flesh nor touch bone. She's still a young one, her skin supple and forgiving. Jeremy squats down and starts on the front paws. Serra screws up her nose as she attempts to find purchase around the fluffy thick skin around the anus without, god forbid, piercing the cavity.

'Can't smell anything much yet,' I comment. My dad is prone to exaggeration.

'Snare-trapped?' says Jeremy, taking note of the gash around the neck. 'Your doing?'

'Yes, snare-trapped, and no, not my doing,' I answer ambiguously. Jeremy raises his eyebrows. 'Let's just call him fox man,' I say, smiling to myself. Serra pauses her grisly task to give me a quizzical look.

Before he became fox man, Luciano was the hipster I met at a fiftieth. I first noticed his offering to the potluck table. Rather than the ubiquitous dhal, quinoa pumpkin salad, or plastic dips and chips, Luciano set down a plate of homecooked silver whiting on a bed of greens. They were perfectly cooked, melting in my mouth. I got chatting to the chef about backyard farming. He told me how a fox had been terrorising his chickens.

'It's only a matter of time,' he said, having experienced previous chicken carnage at the paws of neighbourhood foxes. 'This time, I've decided to go on the offensive.'

'What — catch a fox?' I said with no small amount of disbelief. With a slight Italian accent, cultivated stubble, and knowledge of the correct way to splay pork sausages, Luciano didn't strike me as a hunter. I responded with what I assumed to be equal party hubris, claiming that if said fox was trapped, I would tan its hide and make a winter coat. It wasn't entirely hubris. I had been fantasising about making my own animal-hide clothing again. I had a variety of op-shopped fur vests, all fake, and had been dreaming of wearing something that didn't feel like plastic — something that I made myself — and mulling over where I might get some rabbit skins. Still, I didn't see any city fox pelts coming my way anytime soon.

When I retold the story to my housemates after the party, they pointed out, with great laughter, the flirty connotations of the contract involving trapping, skinning, and tanning. What kind of a deal had I made exactly? None that would come to fruition. That I knew for sure. A fox is a wily creature and is not about to waltz blindly into a hipster's string. Traditional traps take skill, patience, and practice.

A month later, I received a text.

Hi Claire, hope the NY finds u well. We had a couple of conversations, one of fish (haven't forgotten) also fox skins. Well I've set up some fox traps 2day! Of course it's going to be REALLY easy to catch them as u tube has taught me everything there is to know! 😆 *So if i do catch a fox u want me to keep them alive or otherwise? Assuming u are still interested?*

Luciano 🦊

I was surprised and a little intrigued. Yes, I am interested in both foxes and flirting. And I imagined there might be more of one than the other.

Where have you set them? I replied.

Set in my backyard which sits on the creek, so together with chooks at my place foxes visit everyday. If I catch one not sure how to kill it without damaging the hide, any ideas as my traps are snares not a cage.

The question still seemed rhetorical. I still couldn't believe any animal was going to be harmed in the process of our flirty conversation starter. I let him know that a snare trap would choke an animal, therefore not piercing the hide. I suggested camouflaging his scent.

I expect they'd b a bit wary of fresh human scent on their well-travelled path for a little bit. Could buy fox urine like u can in the US.

Wow, he'd really been doing his homework. But to ward off any expectations, Luciano added another couple of disclaimers.

Plus the fox I saw recently had Mange so not what you'd want in a hide.

Indeed it wasn't. A week later, while out bush, I switched on my phone to find an SOS had been sent overnight.

Umm a bit late to text but i just snared a live FOX!! WTF do i do!!! It's alive and making a small racket (hope neighbours don't hear it) too late & in a dark corner of yard to try killing it and not sure how anyway. Will get up early & deal with it if it hasn't got loose. Call me at daybreak if yr around. P.S. it doesn't have mange 🦊

WTF indeed. I couldn't quite believe it. I texted back straightaway, saying I was out of town and asking for more information. I could sense his exhaustion. It was a hot day. He could put the fox on ice. I suggested the freezer. As soon as I pressed send, I could feel him recoiling. The last thing a foodie hipster wanted was a stinky fox in his freezer. I was still hours away and exhausted from running a week-long camp for sixty kids and their parents. I made a commitment to come over at 4.00 pm for skinning and a mojito, but as the morning progressed, my tiredness took over and the idea of combining mojitos, foxiness, and blasting heat was too much. I felt terrible but told Luciano to bury it.

I went over a few days later to celebrate and commiserate. His architecturally designed bachelor pad was more like a castle, immaculate and stylish. A bathtub on wheels nestled next to a tropical garden; the tap handles were made of stones, Balinese-style. His avocadoes were lovingly netted, the tomatoes burgeoning.

Over the back fence was Merri Creek. I met his well-loved chooks and heard the story of the chicken auction where he procured them. The golden layers that had won his affection had also been the object of desire for a ten-year-old lad. Luciano hadn't had the heart to outbid him, so he'd waited for the leftovers, which had proven to be stellar on the egg front. And it seemed, for fox bait.

Luciano led me to the scene of the crime down in the back corner of the yard. I half-expected to see chalk outlines. The broken snare wire hung loose, the metal stopper misshapen from the struggle. Luciano showed me how he had propped the wire loop up at head height and tucked grass and leaves around the outside to channel the fox in. I was impressed. The man really inhabited his backyard, I thought, so much so that he knew exactly which direction the fox would be sneaking. Intimacy with your environment is the only way to outfox a fox. Maybe that and some beginner's luck.

Looking at the setup, it dawned on me that this snare would not provide a short choking death. I thought back to his text — 'It's making a racket in the backyard.' Oh god, I realised with horror, he would have had to kill it by hand!

'Ummm ... I realise I never asked you how you killed it,' I said guiltily, imagining myself in the same situation.

'I know!' The absence of the question had obviously not escaped him. 'I didn't know what to do. I realised that I couldn't leave it until morning, as it was so noisy and there would likely be animal rights protestors out the front of my house.'

My mind raced for solutions. Gun? No, he wouldn't have one, and that really would have alerted the police. Throat slitting? Surely not.

'So ... how?'

In the middle of the night, Luciano had paced around the living room, racking his brains for a solution as he listened to the sound of the fox flaying himself around in the corner of the backyard.

'All of a sudden, it came to me — an image of the old dog catchers with their pole and noose.'

Once again, YouTube had come to his rescue — that and some fast-thinking engineering genius. In midnight's dark hour, Luciano had fashioned a killing device out of a PVC pipe with rope fed through it and fashioned into a noose at one end. Once around the fox's head, a pull on the rope drew the fox into a stranglehold that quickly cut off the airways. Still, I winced. Killing — it's messy business without the buffer of a trigger or sterilised supermarket packaging. I watched him closely as he replayed the story. It wasn't exactly joy, but there was a new aliveness in this intimate interface between land and beast and human ingenuity. I got the sense this fox challenge was as much about waking up some part of him as it was about protecting his chickens. Something that dinner-party conversation couldn't touch the sides of.

A couple of months later, however, after I had just farewelled the last of the 150 people that had come through our backyard for our latest Samhain ceremony, I received another text. It was 3.00 am.

I hope the start to your evening of the dead is going well. Going well here as I have caught that 'winter coat' ready for you to do anything with if you're still interested?

He'd done it again! Another one! Again, though, the timing. I hadn't really slept. One hundred and fifty people had just come through my backyard. The thought of wielding sharp implements on an animal was about the last thing I could handle. Also, we'd been spending a lot of time together, and it was beginning to feel a bit like a game of hide-and-seek. I had been thinking of taking some space. All things considered, I texted him back with a thanks but no thanks, feeling not a small bit guilty, and invited him to consider a second burial in his backyard instead. As I headed out to breakfast with friends, it occurred to me that at just about the time that this fox had been heaving its last breaths, I'd been in a day-of-the-dead ceremony, chanting a song about moons and eagles and … foxes. A little uncanny. The image of a dead fox and the soft pelt of my winter coat lingered over my cafe crepes.

Another text came through in the afternoon.

Last thought, I'm going out 2nite I could drop at yr place where upon you can play with it 2morrow? Yes or goes to tree heavn?

I'm not sure whose snare was baited, but at this message I called him.

'It's in the DNA,' he said cryptically.

'Is it now?' I replied, pretending not to understand the reference but aware that a no would be an afront to both our party banter and some other ancient agreement.

An hour later, there was a hunter at my cave entrance with hessian sack in hand. I opened the door. He was wearing a gender-neutral black and white striped top, red neck scarf, and a hint of eyeliner and held the sack out to me with one eyebrow raised. His face denoted a questioning, a hint of vulnerability and satisfaction. His part of the deal was done. I'd reneged once already on our agreement and now it was time to make good.

'So is this is where we start grunting?' I said, smiling, and took the bag from him. It was heavier than I expected, and I locked my elbow into my hip for support. 'Follow me to my cave.'

In the laundry, we squatted to peer into the bag. The fox was small, curled in the fetal position. It reminded me of skeletal remains of four-legged creatures exhumed from peat bogs. My stomach flipped at the rough bloody gash across its neck. Instinctively, my hand went in to pat the soft hair on its head, as if reassurance was never too late. The pelt was soft as down, grey-brown and flecked with strawberry blonde darkening on the tail.

'A vixen' I said, meeting dark eyes.

'Indeed,' he replied, and I looked down quickly.

'Well, the deal has been upheld, and I shall attend to the beast on the morrow,' I said. 'Even though there does seem to be something missing from the equation.' Luciano grinned as he slipped out into the shadows.

That night, I woke thirsty and half-sleepwalked into the laundry for a drink. The tiles gripped my toes with cold. Suddenly, I tripped on something not part of my spatial mind map. It was stiff and yet soft, and for a moment I couldn't think what it could be. And then I remembered. I had just kicked the vixen. I cringed and gingerly stepped around her to get to the tap, silently apologising. In that moment, I heard a yelp from down near the river. Dog? It yelped again. I paused and kept listening. It was no dog. It had to be a fox pup. I pictured the young tan pup down by the boundary fence where the trail of compressed grass snuck into the bushes. An easy snare spot, Luciano had pointed out, and then noted me bristling at the idea. Why not in my backyard but okay in his? Why did I not want fox blood on my hands?

Luciano requested that he remain an anonymous hunter, so I give little away to Serra and Jeremy, who seem content to focus on the task at hand.

'Got any screwdrivers?' Serra asks, moving from bum to tail. Serra has also been YouTubing, having recently moved to the country, and is intent on learning to skin roadkill. She's about to show me a new trick. I watch as she lashes the screwdrivers together and then proceeds to separate the thick tail by running the screwdrivers under the skin. Halfway down the tail, the makeshift tool snaps the skin in two, and the once-glorious bush is left looking a bit like a stalk with a tuft on the end. There goes my fancy stole.

With three of us on the job, all the major cuts are done relatively easefully. It's time to undress her completely. I take hold of the skin from the top and bring all my weight to bear, with Jeremy straining on the other side. The pelt resists, and I turn my face away, half-expecting the carcass to split in two and spray blood all over us. With a wet, juicy slurp, it gives a little, and then a little more, and then finally peels back like a sticky wetsuit.

With every section, layer, and tuft we removed, the fox gradually lost more of its animal essence. Now stripped of its cloth, it's more carcass than creature — lean, red-muscled meat with an ugg boot on each paw and a coiffured poodle's ridiculous tail.

'Let's get this thing off before the neighbours see,' I say, suddenly uneasy about the repercussions if Rusty the yappy dog fails to alert me to his owner's homecoming. I'm still not sure whose side he's on.

Down at the river, Serra and I rig up a scraping beam by lashing together a rough tripod and leaning a long piece of treated pine across it. With a hessian bag between me and the hide, I put my weight against the beam with my belly to anchor the hide in place as I begin to apply a scraper to coax back the fat and epidermis. The layers peel off with relative ease, especially compared to a deer or goat, which can take days.

'The start of my winter coat,' I say cheerfully.

Serra is like an excited dingo pup scratching around for more fun. She fixes on the idea of extracting the brains to use for the tanning process, already aware of the bizarre equation that every animal has enough brains to tan its own hide. It's something that I was willing to let slide in lieu of some sheep brains from a halal butcher. But in a matter of minutes, she's in there with the fox head on the block, splitter in hand, and tail wagging before realising what she's signed up for.

'Eeeiioow!' Serra squeals as the splitter bounces off the head like it's rubber.

The skull is not an egg. It is one tough nugget, remaining loyal to its purpose of protecting the control centre long after its need has expired. I turn away and keep scraping, my imagination filling in the gaps as the splitter comes down with a series of hollow thuds and attendant squeals. Jeremy stands by as a coach offering encouraging suggestions. Finally, the skull gives way.

'Oooh, grooosssss,' Serra says as she dives her fingers into the cavity to dig out the sack of brains, which she puts in a scat-sized pile on a piece of bark. The poor animal looks tortured now. It's time to put her to rest.

'You're not going to eat it?' Jeremy asks.

'City carnivore? No thanks,' I reply. And I've heard foxes can contain a parasite. Google is ambiguous on that fact, but I'm not going there.

'What do you think?' I say, holding up the pelt, which now looks a lot more like apparel than it did an hour ago. It's beautiful, flecked through with sparks of gold.

Applying the brains to complete the tanning process requires a few more hours, which I don't have, so I roll the pelt up and stick it in the freezer and the brains in a jar in the fridge. I label the jar with my initials; I'd hate for my flatmates to think it's pâté. The carcass goes into the long-term compost, and I cover it with a good load of dry leaves.

I'm a bit sad I didn't accept the first fox now. My winter coat would be halfway there. And maybe that's the last I'll hear from fox man now the exchange has been made.

After a good scrub, I sit out on the deck with a cup of tea and message Luciano photos. From his cave comes a string of paternalistically proud replies.

OMG!!!! U Champion. All of u! So professional looking! I'm so glad I persisted pushing this onto you.

The caveman is pleased. I glow with cavewoman pride.

I'll keep supply coming. Just re-set trap together with found half eaten dove as bait. Any pre-orders — easy to adjust snare?

A young male please, I reply cheekily. *My winter coat needs a complete polarity.*

A few nights later, Luciano comes over 'to soak in some moon rays'. We light the fire down by the river and sip on the bottle of Tuscan red that he smuggled in under his jacket.

'I've got a new strategy,' he tells me. 'I've been practising distressed rabbit calls to lure the foxes in.'

I fall back on the ground laughing.

'Another YouTube find?'

'Of course! It's my guru!'

I like people who can walk in different worlds. There's a wholeness in being able to shapeshift from fox hunter to fine-wine connoisseur. It's a quality I appreciate in myself too.

'Okay, well, let's hear it then,' I say.

He walks closer to the river, his back to the fire, and waits as if listening. In his usual blacks, he almost fades into the shadows. A sound unlike anything I've heard from a human finally emerges, a kind of pitiful whistle, a cry for help with the knowing there's no help coming. It shifts me into a quiet attending, my senses leaving their fixation on the issuer of the sound and travelling down the riverbank. I turn my head, thinking I hear quiet footsteps nearby. Two bright-green eyes peer back at me from the bushes.

'Look!' I hiss to Luciano, but when he follows my glance, the eyes have disappeared.

I can't imagine setting traps for these foxes. I feel too much rapport with them now. But that's also exactly what traditional hunters rely on — intimate knowledge and empathy with their prey. To love and to take life in the same breath. Maybe that's what they mean by the heart of a hunter. If so, it's a strong one.

'You know, when I was a kid, I wanted to be Les Hiddins,' Luciano says.

'Really?' I say, having trouble putting the image of the 1980s TV hit *The Bush Tucker Man* together with this urban native.

I tell him about the conversation about the ethics of urban fox culling that I had with friends over breakfast the other morning as I deliberated whether or not to accept fox #2. We all acknowledged that, rational or otherwise, there is comparative unease at the concept of knocking off a red fox in the city compared with the country.

Luciano leans back on the picnic rug on one arm.

'Well, it's the largest undomesticated animal that we see in the city,' Luciano says. 'The largest wild creature. There's something in that.'

'Totally,' I say, surprised that it's something he's already considered. It's something I've been thinking about the last few weeks. A creature is a product of their environment. How I encounter fox, what meaning and relationship I attest to it, is dependent on context. I'm used to seeing them in wild or rural landscapes, so the urban fox, which is in much higher densities than in the bush, is a new acquaintance. Like Luciano, I've been considering how the sight of a fox in my backyard elicits a sense of the wild, the untamed and uncivilised. In rural areas, foxes threaten wildness, whereas here they seem to uphold it, despite the fact they are still preying on native wildlife.

'Maybe there's some kind of solidarity I feel towards them, both of us in some way protecting or symbolising the self-willed quality of wild places and wild beings,' I say, thinking out loud.

'A protector of the wilds, hey?' he says with a half-smile. This hunter is half fox himself, I think. Equally sensitive, equally cagey. Perhaps it takes one to know one.

My relationship with foxes at the moment feels just as grey.

My childhood was filled with stories of 'fiery-eyed foxes', as my uncle called them. I would hear them yelping from my bedroom at night and clutch the doona around my chin. With my brothers, I would sometimes visit their multi-entrance dens in the back paddock and look for tracks. Explicitly, the family attitude towards foxes when I was young was one of revulsion — and yet implicitly, it was one of fascination and respect.

When I became a conservationist, the image of the fox was drawn from science — an opportunistic invasive species, the main culprit for Australia's record for the world's worst mammal-extinction rate, and a continuing threat for ground-nesting birds, reptiles, and other creatures. This was undeniable.

When I began studying nature observation and tracking, the picture began to get more complex. Once, about fifteen years ago, while on a solo hike high up in a wilderness area, having not seen another human being for two weeks, I sat on the edge of the trail one morning and witnessed a red fox trot past, not metres away, completely oblivious to my presence. My gaze followed its silent form hungrily. Head held high, footfalls silent, pelt lit up like red sparks in the early rays of sun. I was besotted, entranced. It was both graceful and fierce. Something stirred within me. I was allured by its self-containment, its poise.

My continuing fox encounters during this period of my life seemed to tap into a more ancestral or universal relationship with fox. I discovered that they are deeply embedded in our collective psyche, a pan-cultural figure of significance. In Celtic mythology, it's understood the fox knows the woods intimately and is to be relied on as a guide to the spirit world and as a keeper of wisdom. In early Mesopotamian mythology, the fox is one of the sacred animals of the goddess Ninhursag. In Chinese, Japanese, and Korean folklores, foxes are powerful spirits, known for their highly mischievous and cunning nature, often taking on the form of female humans to seduce men. They are knowledge holders, keepers of the silence, edgewalkers, survivors.

In the contemporary urban context, fox mirrors our lost wild spirit. Fox lives in the shadows, the unseen and forgotten-about places that remind us at some level of our own unexplored lives, the ones we are both fearful of and attracted towards. Beneath, behind, and barely out of sight of the millions of city dwellers in every neighbourhood is this wild one, butting right up against our shodden plodding and plastic wrapping. Its invisibility is a particular kind of presence itself. In a direct and immediate juxtaposition to the fences, backyards, and bricked-up castles of our making, fox operates outside of the rules, beneath the radar, beyond the limitations of borders and boundaries. While we feed off each other in a transactional, city-centred loop, the fox uses us but does not need us, is neither at the top nor bottom of the food chain but outside of it altogether, offering the possibility of living on the merits of skill, wit, and instinct, with a quiet confidence born of freedom and self-reliance.

And now my ambitions of clothing myself in fox has opened up a new conversation with the animal. A winter coat is never just a winter coat. When one becomes complicit in the life-death-life cycle, it's about as far from 'off the rack' as possible. It's a courtship with the animal and its familial order, with the niche it holds both in the physical ecology and the collective ecological imagination. Right now, I feel like I'm in multiple compelling stories — the modern and the ancient, the rural and the urban.

Luciano shifts around and looks over his shoulder.

'And perhaps I'm also encouraging wildness by encouraging your inner hunter to emerge,' I say.

'Hunter? Not me. I'm just a humble urban farmer,' says Luciano with a grin. I don't believe him, and I don't think he believes himself. While this man is protecting his domestic chickens and keeping a wild woman in pelts, by entering into the ancient conversation between 'hunter' and 'hunted' he's also putting flesh on the bones of an unlived part of him.

'Wildness is what I kill and eat,' proclaimed the father of the field of human ecology, Paul Shepard, 'because I too am wild.'

The hunter is an emaciated figure in our culture, despite the fact humans have been preying on large animals for two million years. That specific dialogue has shaped us — the ingenuity required to hunt in cooperation was most likely the catalyst of our species' emergence. According to nature writer David Petersen, hunting also advanced us spiritually: the same wild animals we preyed upon, we also prayed to. 'With life and death a daily yin and yang, and the co-dependence of all living things as starkly obvious as the blood on one's hands after a successful hunt, it was utterly natural to view sacred powers as diffused throughout nature.' In these ways, the hunting-gathering lifeway has been a defining becoming-human facilitator.

Western culture spoonfeeds us like a parental bird and yet the hunter instinct is merely smothered, not dead. By consulting his online teachers, reading track and sign in his immediate environment, and employing curious attentiveness to fine-tune his tools and weapons, Luciano *is* edging towards the role of the hunter, a protector of a different kind.

'Okay, well, can you just pretend you're a hunter while I share a campfire tale with you?' I say.

'Only if it doesn't require me to paint my face in charcoal,' Luciano says, stoking up the fire.

'I can't guarantee that,' I say laughing. 'But I'll take that as a yes.

'Once upon a time, there was a hunter who lived in the woods, lonely and resigned to his fate. It was the end of the day, and his boots were wet, and the dusk birds had begun to call, and he was on the familiar trail back to his hut when something stopped him in his tracks — a plume of smoke coming from the chimney. He felt a little scared and wondered who on earth it could be in his home. When he peaked inside, there was a delicious stew on the stove, his raggedy clothes had been mended, and the whole place had been spring-cleaned. An unfamiliar warmth enveloped his heart. No one had ever been kind to him before. All week this went on, until finally on Friday, he had the idea to come home early and see who it was. When he peered in, there was a woman at the table, her long hair flecked red as fire, and she was singing some half-forgotten language. He knew that this woman was part fox, part woman, and part spirit of the forest.'

Luciano takes a sip of wine and looks dubious. I pause for a sip too, and a fox yelps from somewhere upriver. Both our heads swivel to peer into the darkness. We turn back to the flames, and I continue.

'The fox woman turned to the hunter and announced simply, "I will be the woman of this hut," and let it be known a very sweet night they had. And so fox woman moved in, and every night she would hang her pelt behind the bedroom door, and the hut would fill with its wild, regal scent. Now after a few weeks of domestic bliss, the smell started to get to the hunter, and one night he said to the fox woman, "Darling heart, I've been waiting for you my entire life, and this love nest we've created is really sweet … but there's just one thing — it's the pelt — it smells awfully sour. Might you leave it outside from now on?" The fox woman looked at him without saying a word. A few more weeks passed, and the smell of the pelt seems to grow stronger, infiltrating the hunter's mind, driving out all the affection he once had, until one evening he looked into fox woman's red-brown almond eyes with a stony gaze and simply said, "Get rid of it." And in the morning when he woke, the fox woman was gone, the pelt was gone, and the scent was gone. And they say and say truly that to this day, the hunter stands lonely in his whole body for the scent of the fox woman and the haunting melody of her strange forest songs.'

Luciano lies back on the rug and sighs. 'True as myth, hey?'

A friend texts me the next day about a dream she had involving me and a fox. Another good reminder. It's never just a pelt, never just a carcass; it's an entering into the mythological narrative of the species itself.

Gmorning Claire! Feel like doing another pelt? I receive a text not even days later.

No way, I can't believe it! I reply. *You are one alpha hipster.* I chuckle with my provocation.

Alpha Hipster — that's insulting!! … but accurate perhaps. Up since 3am — signed, tired alpha dude.

'A young male, as requested,' Luciano says proudly, opening the hessian bag. Always the moment of shock, taking in the sight of the stiff, still body, which only yesterday turned its face to the morning sun. The pelt is a little less full and soft, this fox obviously having been around the block a few times.

The story of the hunt must be told. Luciano takes me to the crime scene. Evidence number one, a ripped shard of wire hanging from the neighbour's fence into the park. Evidence number two, a star picket half-covered in dirt.

'What happened here, tracker?' he asks.

'No — it couldn't possibly pull the star picket free?' I say incredulously, struggling to imagine how an animal that size could be that strong.

'Yep, and then dragged itself to the top of the neighbour's fence with the star picket, which promptly got stuck. Imagine if I hadn't heard it and the neighbour came out in the morning to find the fox hanging off its fence like some pagan sacrifice,'

Luciano says. 'Instead, it was me at 3.00 am in my dressing gown, lasso in hand.' He gestures the final scene. It's a charade of black humour.

Luciano's snare technology has been significantly upscaled since the last event, this time involving fish sinkers and minute height adjustments following mathematical equations related to the relative distance from the ground of the head of a cat compared to that of a fox.

'It's 100 per cent cat-proof,' he says, jokingly wrapping the apparatus around my neck.

There's something curious going on here with this foxy business. I had just called for some space between us when not one but *two* foxes jump into his traps within the space of days. Who is trying to trap who here? Whose cunning is this really?

I text around for my next skinning apprentices. My friend Michael is keen. I almost tripped over myself when someone told me that Michael was learning how to hunt, since I knew him as an apartment-dwelling, latte-sipping type that I enjoy dancing with from time to time. With the hope of being able to hunt my own wild meat, I recently bought a compound bow and have been accompanying Michael to an archery range to practise.

I'm feeling more confident about skinning this time, laughing with Michael as we tie the fox up to the same branch, now accustomed to the weight. He unfurls his burgeoning knife collection, and I pick up a small flip knife to try out.

'First time?' I ask him.

'Yep, I'm a skinning virgin. Show me the ropes,' he says enthusiastically.

We're halfway through the hide when Luciano appears on the back path. It's a one-sided hug he gives me as I keep my bloody hands to myself. The pelt has more of a scent this time. It's the smell of fear. He stands back with arms folded.

'This is not a spectator sport,' I gibe and throw Luciano the soft, fluffy testicle sack.

'Hey, I don't need any extra masculinity, thanks,' he says.

This time, the skin rolls off like a sock. I'm intending to tan this one with eggs and oil. Hearing my desire to bury and then exhume the skull, Luciano takes it upon himself to separate it from the body. It's harder than he thought, and he's wrist-deep in blood and grimacing before it finally gives way. I wince, and my stomach churns.

We're so removed from the blood-and-guts reality of life and death. Last week, Luciano and I killed and plucked roosters, and even that was challenging. He had bought three at a chicken auction. It took a little getting my head around the facts and motivations of purchasing and transporting three cocks past their use-by date (for $5 a pop), knowing that he is not short of the cash to buy an organic chook straight off the shelf. He explained. These were 'Light Sussex' heirloom variety. They don't have the big thighs and fat belly that we have become accustomed to. They don't make it into our hotpots or ovens anymore. At best, they make it to the petfood production line. So, not only would we be eating something that would otherwise be thrown away (therefore saving the chicken we would otherwise buy), we were encouraging the breeders to keep the old stock alive. It's a form of saving seed — keeping the funky old varieties in circulation.

He handed me the first cock and indicated to hold it upside down by the feet. 'Don't let go,' he instructed. 'Blessed are you and those who are about to eat you,' he said as his mother had taught him. I turned my head away as he drew the knife across its throat. The chicken struggled in my hands for a couple of minutes, slowly going quiet. I hung it up on the rope line to bleed, a bucket underneath catching the dark-red rivulets that slowed to a trickle. Next, we dunked each one in a pot of boiling water and promptly got plucking, feathers clinging to my hands as if in honey. Luciano packed one bird up for me to take home, instructing me to put the whole chicken in a bowl of water overnight with two cut lemons to soften it.

Luciano is definitely not one for sentimentality. At the fox fleshing beam, we take globs of the gooey fat from the underside of the skin and rub it on our leather boots until they shine. My freezer is growing fat on the largesse this dark prince procures.

'One more pelt and I'll have my winter coat,' I tell Luciano.

'Let's see what the fox drags in,' he replies.

I don't have to wait long. I take the third pelt notification when out of town. I text my housemates to tell them Luciano is going to pick me up from the airport. *Another fox call?* Megan messages back with a laughing emoticon. It's become the running joke of the household.

'Who needs flowers?' I say wryly as Luciano opens up the boot and squeezes my luggage next to the now-familiar hessian sack. Once home, Luciano spills the contents out onto the path. It's another vixen, smaller than the first. She's deep in rigor mortis. I pat her pelt, a little teary.

Steph, one of the Rewild Friday crew, volunteers to help me undress this one. She's slow and methodical, and I'm grateful for the quiet company. We chat about her plan to spend a month alone in the bush. At some point, my thoughts get ahead of me, and I cut too deep around one paw. The tendon snaps, and the paw falls limply.

I hear a noise out the back. It's the plumber come to unblock the drain. Although half-hidden by the trunk of the dogwood, I register a flash of confusion on his face. He must have seen us. I tuck my knife into my belt and smear the blood from my hands on my pants. What story might he tell himself to make sense of this strange scene, I wonder. A dead animal strung up in the backyard of a well-to-do suburb being taken to with knives by two chicks in boots. I bury the fox near the outdoor fire-heated bath. The backyard is beginning to be a bit like a pet cemetery.

The following weekend, at the encouragement of my housemates, I remove hide number one from the freezer, whizz up the brains in the blender, and begin smearing the resulting liquid over the flesh side of the hide. Sparking up a small fire, I sit on a stump and work the skin with my hands — stretching, pulling, abrading over rocks — opening up the fibres and coaxing the natural chemicals in deeper. The skin soaks it up thirstily, becoming increasingly supple and flexible. My fingers soon tire, but I know that if I stop before the hide is completely dry, it will be stiff in sections, and I'm aiming for skin-on-skin wear.

As the smoke passes over, I contemplate the events of the last few months. There's many different versions of the story I could tell. Many that others could tell too. Perhaps just as many one might tell about fox itself. Probably all true in part. That's part of the shapeshifting beauty of both fox and myth alike, slippery and subtle, sidestepping traps of black and white definition and singularity. I hold the pelt up to my chest and stroke its downy softness, imagining my winter coat.

Wrapping myself in fox furs might allow me to more easily and more often step into the shoes of the fox woman — to entice the ones in from the cold who have forgotten the warm fires of company, to help them remember the beauty of the wild pelt that they too once wore. Somewhere in history, we have exiled the fox woman and her untameable life-death-infused pelt, have banished the wild soul from our homes and hearts. I'm ready to welcome her back to my hearth, perfumed even so with that unfamiliar woodsy scent, this time with new tales on her tongue to tell.

Epilogue
Living in the New Normal
2020 and after

'How are there no mushrooms at all?' I say, squatting down to look under the mat of pine needles on the forest floor. Scattered around me are shreds of decapitated fungus, yellowing white against the dark brown of the decaying needles. 'Looks like a trawler's been through.'

'Hmmm, yeah, it does look a bit trashed,' says Taj, still optimistically clutching her large empty cane basket. A grey shrike-thrush echoes out a single bell-like chime in the otherwise eerily quiet pine forest. A breeze whistles softly through needles up high.

'You look like someone from a movie set,' I say, pausing to take in Taj's long, flowing maroon dress, black monster boots, Mediterranean skin, mass of curly blonde hair, and knife hanging from her belt.

'Maybe the working title is "Starving in the Pines",' she jokes, and I laugh, my belly rumbling too. This mushroom mission isn't a tourist daytrip: we're counting on it for food. It did occur to me when driving out here that the Monday after the first sunny autumn weekend since the lifting of COVID restrictions could be about the worst forager's plan, with every cabin-fevered family looking for a wholesome outing close to the city. Seems like I was right — we haven't found a single saffron milk cap.

'Look! Slippery jacks!' Taj calls out excitedly. The brown and slimy tops of a ring of the young and edible slippery-jack mushrooms are barely visible beneath the hat of needles they've pushed up from the forest floor. A small search in the same area reveals another few rings, perhaps too camouflaged for the weekend hordes to find. We harvest a dozen, leaving the youngest ones, and wander further to collect another half-dozen.

Taj pulls a thermos and two ceramic mugs out of her bag, and we nestle in with our backs against the hard shingles of the pine tree so that our shoulders are touching. The steaming liquid smells like something I imagine emanating from an underground apothecary. I take a sip. It's sweet and so earthy it's like Taj has nipped off a root and ground it up in my mug. My hunger pains are quelled instantly.

'Roasted acorn, bay leaf, chicory root, and honey from your bees. Pretty good, hey?' Taj says.

'It's ridiculously good,' I say, wrapping both hands around the warm vessel. We both pause as a rabbit hops by on the other side of a ditch.

'Hey, did I tell you a woman got fined $1,600 for foraging alone out here at Easter?' I say.

'What? Are you serious? But foraging is an essential food mission.'

I was shocked too when I heard about it. It seemed punitive rather than protective. It made me wonder about the real motivation behind the policing of the foraging grounds, especially considering the proposed local by-laws I heard Patrick and his partner Meg had been fighting last year that threatened to ban foraging, lighting fires on public land, repurposing materials from the tip, planting fruit trees on nature strips, collecting firewood from public land, even setting up roadside stalls. But this was worse — these were basic human rights! Taking away the right to forage would be akin to banning kissing! Could lockdown tame us into a new normal even more controlled than before?

'Well, foraging is certainly is an essential activity for us right now,' I say. We look down into our scarce baskets.

'Definitely not the week's worth we were counting on,' Taj says.

'Such is the forager's fortune,' I say, finishing the last drops of tea. 'Let's just hope our fate shifts, otherwise it's going to be a hungry week.'

Today is day one of what Taj and I have coined the Urban Hunter Gardener Challenge — a self-designed week of eating only what we have foraged or grown, or that friends gift or barter us from their own grown or foraged food. It's part treasure-hunt-adventure, part fun, part philosophy-in-action. I want to see if I can successfully unplug my umbilical cord from the monetary economy and plug it instead into a direct exchange with the earth. Do I really have the knowledge, skill, and ingenuity to pull this off? Can the city's largesse be fruitful enough to feed me without cash in my pocket? The greater the need, the greater the result, my tracking mentor Tom Brown Jr would say; if you want something badly enough, create the ultimate need for it. Hunger's a pretty powerful motivator. I'm hoping my growling belly will guide me into greater knowledge of this urban land I've been befriending, and show me that the city is also a part of the tree of life and not just a consumer of its fruit.

I'm glad Taj accepted my invitation to join. The original plan was a two-week challenge in April, timed for acorn season. This meant a mere two months of preparation — not very long considering carbohydrate vegetables often need more like three months to grow and my pantry was empty of preserves. I got busy and planted veggie seedlings, roused Luciano into helping me build a chook pen, foraged and preserved summer plums, blackberries, and grapes, made a trip to the bay to forage seaweeds, experimented with my first homegrown zucchini and cucumber pickles, and put feelers out for animal-protein possibilities.

When COVID locked the city down in March, we reluctantly acknowledged that our free-ranging foraging plans were incompatible with the current limits on movement. We wouldn't even be able to swap veggies with neighbours without negotiations and sanitiser. We shelved both the preserves and the plans for the foreseeable future. Now, with movement allowed once more, we've dusted off the plans but reduced our ambitions to seven days.

It's already been a great disruptor of routine. Instead of ruminating in bed this morning, I turned my mind to a single focus. I sprang into my walking shoes and with knife and collecting bags took to my usual route around the wetlands. Just shy of the trail, I found a tree dripping with red-ripe kangaroo apples. How had it sprung up overnight? I couldn't believe I hadn't noticed it before. Edibles jumped out at me from every twist in the path. There were potatoes and artichokes growing wild in an abandoned veggie garden, dock seeds ready for harvest, sweet fennel seeds for tea, a thick patch of young mallow that I picked for my morning omelette. Even a single violet that I carefully tucked away for my lunchtime salad. How much more I notice and appreciate about a place when looking through the eyes of a forager.

I had lined up two morning barters. The first was for milk. Google directed me down a suburban cul-de-sac not far from a major shopping centre. Around the side of the house, I found Maria in a red skirt, woollen stockings, and leg warmers surrounded by three chatty and plaited girls. They pottered around a verandah seemingly as large as the house itself, which, Maria explained to me, was home to two families. The backyard overflowed with produce, and at the back was a pen for four goats, who were milked daily. The gate was open to the next house, which I learnt was where Maria's brother-in-law lived. Eight other properties on the street, both rented and owned, were part of the loose cooperative known as 'the hoodie'. Despite being a stone's throw from a major shopping centre, I was in a piece of 'rural suburbia'. Maria handed me a three-litre jar filled to the lid with creamy goat milk, for which I exchanged some passata and a plan to return for a proper cuppa.

Next stop was salt. I rode a kilometre north-west along a dangerously busy and bike-lane-free road to find Kat Lavers, a member of Melbourne's permaculture hall of fame, known for her self-sufficiency genius within the townhouse backyard known as 'The Plummery'. I followed Kat and her single long plait down the hallway and into the kitchen, where I was suddenly transported to a country farmhouse, its open pantry stacked to the ceiling with preserves, its kitchen table overflowing with produce. Little blue-spotted quail eggs rested in a bowl like decorative porcelain. A couple of months ago, I had got a tip-off that Kat was making her yearly salt-harvest pilgrimage to Pink Lake near Dimboola, four hours north-west of Melbourne. The arranged barter was a load of the pine mushrooms I assumed I'd forage and some chicken eggs. Kat handed me a large jar. Inside was a fine salt of the palest of pinks, the colour of desert sunset. She also stuffed my panniers with chokos, grape juice, lime chutney, and a bag of my favourite fruit — persimmon. 'I've been wanting to attempt what you're doing myself for some time,' she said, shrugging off my effusive gratitude. 'So an exchange will be the stories.'

Salt and milk — tick. Now all we need is fat and protein.

A couple of hours later, we've given up on both. Taj decides to stop in at a friend's farm on our way back. The friend and her adult daughter race around on quad bikes, tending to animals. Daughter number two arrives coated in sawdust from an afternoon chain-sawing firewood. The women greet us warmly and happily hand over a large chunk of home-slaughtered lamb in return for a day of labour in spring. As the sun drops, we pick bushels of nettle from their paddock verge. How quickly our foraging fortunes have changed!

'I'm so hungry I could eat a sheep,' Taj jokes as she sets up the gas stove on some stone slabs on the edge of the paddock wind break.

'I'm so hungry I could eat potato chips,' I say, smiling as I produce a few homegrown spuds from my jacket pocket.

'OMG,' says Taj and pulls out her own basket of goodies, containing a sauerkraut from the community garden she's involved in plus some preserved lemons.

Taj drops the lamb chops in the pan, and within the first minute of sizzle my mouth is dripping with desire. Taj finds a piece of tin in the grass and dusts it off to make a lid. I slice up the slippery jacks, potato, and home garlic, throw together a salad of foraged greens and dress it with the apple-cider vinegar I've been brewing with apple scraps from Min's parents' farm.

A dusk breeze blows up. It's often windy around here — the result of the woodlands being cleared for sheep. I make myself into a human wall around the gas ring and wrap the shawl tighter around my shoulders. Taj's chatter gets carried like a dandelion seed on the wind, and I look out to the horizon where the sun tickles the tops of the hills. It's my first trip out of the city since lockdown. I breathe out an involuntarily deep sigh, the kind that comes from a breath held inside for a long time. It's been an intense six weeks, to say the least.

I returned home from a week away and offline to find a world in preparation for the apocalypse, with people buying up toilet paper and pasta by the trolley load. Even the Riverhouse was swept up in the madness, with one housemate spending hundreds of dollars on things like powdered milk, paracetamol, sacks of baking flour, and bags of lentils. Having been cold and flu sick for much of the previous year, Min was understandably terrified of infection. Lockdown would save lives. I was up for that. What I couldn't bend my mind around was this new currency of fear and panic. We battened down the hatches, and placed spray bottles, handwash, and antiseptic wipes at every entrance.

The loss of freedom was swift and shocking. I couldn't see my lover. Our housemate Tully's usual strategies of extreme exercise to keep himself sane were no longer available. The co-working space I'd been working from changed the locks overnight. Gone was the midmorning chai and banter with the same cute barista, the jokes with my workmates. Gone was the swim and sauna on the way back from work, the potlucks and parties. Small losses that cumulatively destabilise your life. I felt like a wild animal suddenly encaged, growling and gnashing at the bars before slowly accepting my fate.

House meetings were full of tears and anger as we struggled to reconcile the new normal that our private lives were no longer ours alone. Who I breathed on was suddenly everyone's business. Share-house harmony had been based on the values of cooperation and community but ultimately placed individual freedom at the top of the pyramid. We'd come and gone as we pleased. Rare was it that the four of us sat down to dinner together. Although we were accountable to each other in ways most share situations don't desire or attempt, lockdown was next level. We were *all* home *all* the time. The kitchen to oneself was a luxury of the past. The old weatherboard house was now a lifeboat floating in a sea of uncertainty, its four inhabitants struggling to paddle in the same direction.

As the days ticked on, we slowly adjusted our privacy settings to this new lockdown intimacy. A whiteboard calendar in the lounge room announced daily events — music jams, Mexican nights, workouts, tax dates, working bees, lounge-room dance parties. We planted gardens. Put finishing touches on the chook pen. Built shelves. Created paths. Cleaned cupboards. Stacked wood for winter. Lit campfires at dusk.

My life was no longer mine alone to navigate, but the joy of being nested within a small tribe was the flip side to the equation. I had three companions, three witnesses to my days and doings. The loss of autonomy gave rise to the satisfaction of interdependence. Life could not be siloed, compartmentalised, or cordoned off from others. There was no 'away' to go to — nowhere to run and nowhere to hide. Entrapment was in equal measure belonging.

Confined to the immediate environs, our little community naturally began to extend out to include our neighbours. Sylvie and her toddler, Dolly, started a scrap bucket for our chooks. I regularly bumped into Eva and her mum in the park with their dog Rusty. We found ourselves saying things like, 'If you need anything, let us know.' I offered to shop for our older neighbour Cas, who continued to drop off dumpstered bread for the chickens. Another neighbour dropped off some foraged field mushrooms from the parkland. One evening by the fire, I looked up to see a dark shape climb over the fence. It was Mike from a few doors down, come to make our acquaintance — with appropriate social distancing, of course. With his partner stranded in Scotland and him working from home, Mike was keen for a yarn. We popped over the following day to check out his garden. I picked watercress from his abandoned pool and a few leftover figs from summer. The world had shrunk in size and deepened in connection.

Lockdown was, as they say, the best of times and the worst of times at the Riverhouse.

'It's ready,' Taj says, and I spin around with sounds of anticipatory glee and help her transfer the chops, mushrooms, and fried potatoes into our bowls.

'Well, almost. You've got to try this too,' she says, dropping some nettle directly into the hot lard. The stems shrivel and stiffen. I pinch one out of the pan.

'Oh my god, nettle chips! So good!'

'*Itadakimasu*,' we say, clinking our bowls together. I eat first with my eyes, taking in the blood red of Taj's sauerkraut, the yellow of the mushrooms, the forest-green nettle, the steaming meat. I wonder if my alms bowl will be this abundant over the coming week. I flip out my pocket knife and slice a thick fatty piece of lamb. It's perfectly cooked, just a faint rosy hue. I bring it to my mouth and lean back against one of the rocks in bliss.

'You can't be serious,' I say, not sure when I've tasted anything quite so insanely good. The whole context is pleasurable: simple food gathered close to its point of origin, cooked and eaten outside when hungry and in good company. Life doesn't get much better.

When I arrive home, there's two bottles of olive oil sitting on the outside table from my friend Anatolia's olive farm. Day one, and I'm being well looked after.

Day three — acorn-processing day — and we've called in a few helping hands to join us around the fire circle and shell the bucketload of acorns I've collected from under the oaks at Studley Park Boathouse.

'These plums are amazing!' Taj says, spooning another of my preserves from jar to mouth. I grab a spoon and follow suit.

'I know, I'm pretty proud of them,' I say, remembering the chaotic mess in the kitchen as I attempted my first mass plum preservation in late summer. Up to my elbows in plum juice and honey, I imagined a moment like this, where the sweet bottled juice would be like ambrosia dripped into my mouth.

'I'd kill for a coffee, though,' she says wistfully.

'Acorn coffee not doing it for you?' I say cheerfully.

My sweet, milky, gingery chai first thing was what I'd imagined missing the most, but in a lightning-bolt moment I remembered my potted cinnamon myrtle in the front yard. I take a handful of leaves turned to aromatic shreds in the coffee grinder and put them into a small pot of goat's milk sweetened with honey. It's continents better than my usual fare.

I bring the base of my water bottle down with a dull crack on another acorn. Removing the shell, I toss the naked nut into the large bowl that sits between Taj and I. Across the fire circle, Cat sits on a rug in front of another stump littered with acorn shells. Ever the inventor, she's experimenting with breaking open several acorns simultaneously.

'Many rotten?' I ask, nervous that the weeks on the ground before collection might have rendered them rat food.

'Maybe … one in six?'

That's a relief. We're counting on these as a staple for the next few days. I attracted a few curious stares as I collected them, most people having no idea that the ubiquitous acorn littering the streets and parks of our suburbs this time of year is one of the city's natural free superfoods. Even in foraging circles, it's not nearly as popular as pine mushrooms.

Acorns are definitely not fast food. After cracking them, we'll be boiling them with about six water changes to remove the tannins. Lucky I'm not too hungry. Brunch today was an omelette care of my chickens with warrigal greens from the garden, pine mushrooms I exchanged for some plums, and Kat Lavers' lime chutney. Yesterday was also egg heavy, supplemented with roasted potkin, an unusual pumpkin variety grown by my uncle.

My friend Sam saunters down to us with a red beret over her long brown curls and smiles broadly at the scene in front of her.

'Hugging?' I ask.

'Only like this,' she says, wrapping her arms around herself. I smile and do the same. Sam has returned to living with her parents while suddenly and unexpectedly separated from her partner in New York. She's apparently one of 400,000 people in the country who for mainly financial reasons have had to retreat back into the parental nest during the pandemic.

'Ahoy there!' a voice calls from the river. It's Elizabeth, pulling up her banana-yellow kayak at the jetty.

'Lady Elizabeth, how wonderful of you to grace us with your presence,' I sing down to her.

''Tis my time-honoured duty. And I come bearing birthday cake,' she says, extending a Tupperware container.

'From which tree was the cake picked, dear?' I joke.

'Oh, I thought you could eat anything gifted to you?'

'Ha ha, that would make it too easy.'

The sight of chocolate cake sparks an instant craving. A few times, I've absentmindedly opened up the fridge before remembering there's nothing there for me. It was only when I caught sight of cheese, butter, and chocolate that I wanted them. Mostly, though, I've felt detached from the contents of the pantry, repelled even when I registered how far some of the items have travelled. Why eat cashews when chestnuts are in season? Why would I buy buckwheat flour from China when acorn flour is at my doorstep? Feeding myself is my primary work this week. Anything else is a bonus. This re-prioritisation feels like a true responsibility, a kind of adulting. This work seems to be taking about four hours a day, which is around the same number I've been told was the daily average for food procurement for Indigenous Australians. In these four hours, I'm also exercising, having time in nature, and often interacting socially. It's life stacking. And of course, I'm not spending a cent on food or the gym, which decreases financial stress.

With stumps arranged at an appropriate social distance for Sam and Elizabeth, so begins another layer of rhythmic thuds. I imagine we look not dissimilar to people from the Ancient Greek or Native American cultures who were amongst those who made bread from this fruit of the oak. Conversation is slow to get going — all of us are a bit socially rusty after six weeks in lockdown. Taj puts the first pot of acorns onto the fire to boil.

'Gawd, I'm missing physical contact,' Sam says with a sigh. 'I'm so ready for a hug.'

'I don't want lockdown to end,' says Elizabeth. 'I've been loving it.' I'm not surprised Liz is thriving in lockdown. Her share house of six people have nested together in style, sharing three meals a day and designing fancy-dress parties with culinary challenges. I've actually not seen Liz this happy in a long time. 'I struggle with too much choice,' she says. 'I got to cook for everyone every day and not have to make decisions. It was great!'

After my initial gnashing at the cage doors, I too settled into a quiet home rhythm. When I crossed out every work engagement in my diary for the foreseeable future, there was an overwhelming sense of relief. I converted the spare room into my creative incubator, with a standing desk looking out onto the tree canopy. For a few weeks, there was barely an appointment to slice up the day. I wrote in the mornings over many cups of tea and emerged in the afternoons to wander and tend to the garden. Sometimes, I would curl up on the futon and just watch the movement of wind through the leaves until the sky darkened and the ringtail descended from its drey. The undefined days blended into each other, taking on a surreal, dreamlike quality. My night dreams grew more vivid, full of powerfully strong and edgy energies that spoke to me of the undercurrent of collective chaos and uncertainty that belied the apparent calm. I had no desire to try to shift my former life online, as others were attempting. That would be missing the point. It seemed to me that we were all being sent to our rooms for a reason. And it wasn't about business as usual. It was for slowing right down. For stillness. Listening. Questioning. Winnowing. Waiting.

Nature too was enjoying the quietude. Whales were able to hear each other in migration for the first time in our lifetimes due to the reduction in low-frequency sound pollution from shipping. Wild-faced mountain lions reclaimed the city streets in Colorado. Without the drone of traffic, I could hear the grey-headed flying-fox colony screech and whinny from my bedroom. Without any correlation I could rationally draw, flocks of yellow-tailed black cockatoos began careening down the Yarra. It was like a call to prayer, and I often found myself rushing out onto the deck to greet their flight. There was a heightened poignancy to their presence that spoke to me of the pathos of the times. The plane-tree leaves on the street began to drop. Maples reddened to bronze. From my comfortable and warm nest, I looked out to a world in crisis and wondered where this was all going.

The devastating bushfires that had ripped through the entire south-east of Australia last summer had already rocked our sense of safety and normality. Climate change was taking some strong strides in our direction, holding up a sign saying, 'No one is immune.' And now a global pandemic. The inherent fragility and uncertainty of life, so easy to forget when encased in the bubble wrap of routine, had been exposed.

There is a meme flying around social media that says brilliantly something I've been mulling over. In the cartoon, the coronavirus is a fish in the ocean being swallowed by the larger fish of economic collapse, which in turn is about to be eaten by the shark of climate change. The COVID crisis surely seems like a dress rehearsal for the larger predator circling. In order to save lives, the unthinkable happened with incredible rapidity — flights were grounded, profits plummeted, we stopped rushing around like mad things. Normal, COVID suggests, is a state far more fluid than we'd been told was possible under capitalism's growth juggernaut. Just as conversations with neighbours opened up in the shared predicament, so too the end of normalcy opened up a broader public discourse on what the new normal could look like.

Charles Eisenstein commented at the time, 'For years, normality has been stretched nearly to its breaking point, a rope pulled tighter and tighter, waiting for a nip of the black swan's beak to snap it in two. Now that the rope has snapped, do we tie its ends back together, or shall we undo its dangling braids still further, to see what we might weave from them?'

As the world unravelled its braids, so too my own neat plaits were coming undone. On the fifth week of lockdown, a trapdoor opened and I fell down into a morass of old griefs and fears. Searing questions of integrity and how I spent my days assailed me. What do I really love? What do I truly value? Am I truly living in alignment with my answers? I switched off my phone, shut my laptop, and let the questions burn through me. One Saturday night, I covered the sweat lodge structure with blankets, heated the rocks, and prayed. This short but deep descent was a microcosm of the larger one at work. In many ways, the collective existential crisis had the hallmarks of initiation. Normality had been suspended into a liminal state of uncertain duration. Initiations are not always successful, though. It depends on the quality of submission, the depth of listening, the willingness to pivot in an entirely different direction. Would this radical discontinuity of life be enough of an interruption to create new patterns and break away from old patterns? I hoped so.

The interconnectedness of our global village has been made more evident than ever. It was a perverse connection to the bush when smoke from the fires hundreds of kilometres away clouded

Melbourne in smog. The boundaries blurred between city and country as we breathed in the ashen particles of koala, possum, kangaroo. The world grieved with us. And then a sore throat that started in China rippled out to affect every corner of the globe. For all our technology and invention, COVID reminds us that we are indeed biological beings, subject to the laws and vicissitudes of nature.

Perhaps this is one of the gifts offered us within this initiation: to let the shaky truth of our vulnerability crack us open so that we might realise our interdependence with all of life. Every action we take on this earth ship ripples out to every other passenger.

After three boilings, the soft acorns get passed around for assessment. I bite off a crumb, the tannins still strong enough to sap the saliva from my tongue. Taj offers to take them home and finish the boiling process. No acorn bread today. That's okay. I have other dinner plans.

Before the lockdown, my parents had visited me for the first time since I moved to Melbourne. We shared time doing what all three of us most love — wandering around, looking at plants, and gardening. They were keen to help me bed down some carbohydrate crops to feed me for the Challenge. Tonight, it's time to dig up those crops. I upturn the potato bags and fish out a couple of dozen small round spuds. Not a lot, but enough. I turf up a few carrot tops and exclaim out loud with joy when they reveal a handful of carrots, fat and straight. With a digging stick, I uproot a large handful of onion weed from the backyard that smells as strong as its namesake. I fry it all up together with a choko and loads of Italian herbs from the garden.

There's one thing missing. Protein. This time, I do turn to the fridge, to retrieve the rabbit I've had marinating in lemon juice overnight. A friend put me onto a guy who hunts them with ferrets in the hills just outside the city. 'Instant death,' he told me as he delivered two lean-looking carcasses in a box to my front door. 'You don't get more sustainable that this.' Luciano recommended the lemon juice, and it's done the trick, the meat falling off the bone and into my hungry mouth. The soup fills me from the outside in. I sigh in satisfaction.

As disappointing as it was to delay this adventure, the timing actually feels perfect. Within the context of dystopian scarcity that stripped the supermarket shelves of flour and toilet paper, the small, simple acts of this week, such as making a bowl of soup, carry more gravitas, a gentle defiance. Rather than shoring up security, as has been the impulse of many in this time of uncertainty, I'm deliberately making myself more vulnerable, turning away from the monetary economy and opening to the economy of the gift, the economy of kindness and generosity. It's an act of hope. Of resilience.

Day four. The Facebook posts of our foraging adventures have attracted hundreds of likes, 'loves', notes of encouragement and inspiration, and — more importantly — food offers. It's time to go foraging in Melbourne's backyards. First stop is a woman called Elly who's offering me bell peppers. Google Maps reveals she's barely a kilometre away. I pull up outside a block of flats, and an attractive dark-haired woman appears in yoga tights and with bubbling enthusiasm ushers me through the garage and into the tiny backyard, where two planter boxes and some pots are the extent of her garden. Elly sees only abundance, however, telling me with the radiance of a gardener in the first bloom of love about her summer crops and the preserves they made.

'They just keep giving,' she says with effervescence. Through the back window, I can see her two primary-school-aged daughters at their lockdown lounge-room desks, seemingly diligent with colours and paper, while her partner washes up at the kitchen sink behind them. 'One has taken to homeschooling with ease, and the other I have to constantly keep motivated,' Elly says with a smile. It looks like a scene of domestic harmony. It reminds me of the 'alchemical huts' Martin Shaw talks of when describing the millions of households incubating together in lockdown in varying states of integration or disintegration. Elly picks me six bell peppers and a large sprig of parsley. One of her daughters comes out to read us the short story she's just written. They wave me goodbye, and I leave stirred by the joy this micro-acreage is gifting Elly and her family.

I head west until I hit Merri Creek and then follow it upstream along the bike path until I find Rachael's house. Herbs and veggies vie for space within the small nature strip and front yard — passionfruit climbing over the last of the zucchinis, sage and salvia drooping with purple spires. A pomegranate tree at the front window hangs heavy with edible red lanterns.

'Coming,' Rachael calls to me in response to my cooee. I can hear the raised voices of kids in the front room. She appears with phone in hand, looking a bit harried. 'I'm completely over homeschooling,' she laughs with a smile that tells me the contrary is equally as true. 'I'd rather be doing what you're doing!' she says. 'It's such an awesome idea!'

'I know, I'm loving it,' I answer honestly, feeling a bit like a kid at Halloween as I bring out my bread bag, ready for goodies.

I was lured by Rachael's offer of rhubarb, imagining a crumble with chokos plus acorn and dock-seed flour. Rachael points me in the direction of the small stems of rhubarb, while she picks me a few pomegranates.

'How's lockdown been for your family?' I ask, the question now ubiquitous.

'For the most part, really good,' she says. 'I reckon we've actually had more nature time than if we had been living our normal busy lives and going away on weekends. We've learnt so much more about the creek just by being around it.'

'Ah, that's interesting, I've heard that from a few people,' I say. Certainly the parks and river trails have been way busier than usual, especially over Easter as the mass exodus was redirected back home. It was the same in summer too as the city became our bushfire bunker in lieu of the usual migration to the coast.

I ran a nature-connection webinar for a local council and asked the ninety participants on a poll whether their nature time had gone up or down since lockdown: 77 per cent indicated that they were spending more time in nature, with 22 per cent 'significantly' so. The chat function flooded with seasonal nature observations and an excitement to have the stories shared. Elizabeth, the outdoor-education facilitator at a local Steiner school, transferred the annual year-nine camp to each child's backyard. They found their own sit spots and reported their stories to each other online the next day. Being grounded, we've been grounding in the reality of place. This is where you are, the invitation suggests. This fox track in the mud. This raven perched on the powerlines. This cricket calling in your long grass. This is the nature that chirps and trots and caws all around you. Come know me as the neighbour that I am.

In her backyard, Rachael picks me a few lemons and some lettuce. I've forgotten the passata for exchange and promise to deliver it later. Rachael shrugs it off and offers me a hug goodbye. Wow — hugging. Tears spring to my eyes at the touch I've been unaccustomed to. This experience of being greeted at the door with fruit and touch is the perfect medicine after the isolation of lockdown. My belly grumbles. I check my phone. Taj has sent me a photo of breakfast — acorn pancakes with poached pears, persimmon, and preserved plums. It's definitely my next stop. I turn the handlebars north-east.

As I chug up a long hill, I reflect on the fact that I'm burning more calories than I've gathered this morning, but I feel full in a different way. I'm nourishing something much more important than my blood-sugar levels — community. I think about the deliveries to my door in the last twenty-four hours. Another bag of persimmons from Emma, another jar of home-pressed olive oil from my friend Claire, a bag of feijoas and perpetual spinach from Rolf and Amanda, a bowl of ripe Irish strawberry fruit care of a spontaneous suburban find by Cat. While I could get by on my own reserves, this project has never been about self-sufficiency. It's about community sufficiency. I couldn't have attempted this five years ago, maybe not even three. The strength of the project is not because I've got a failsafe home food security plan or a stash of two years' worth of preserves but because of the village I've been nurturing over the years through the hundreds of hours of conversations, the commiserations and congratulations offered, the shared fires, walks, projects, potlucks, dance floors, and house moves. Tending to these connections has been more important than tending to the vegetables in my garden. In a way, this project has made visible the interdependent web of relationships that I exist within, and the resilience there within. The community of the city.

I post some photos and stories of my adventures on the Rewild Fridays Facebook group, the community still active online despite the fact we haven't been able to gather this year. I miss seeing their smiling faces. The pandemic has been a shake-up for many of the crew.

Elena finally plucked up the courage to resign from her job and is currently living and working on an organic farm in a community of like-minded volunteers in Byron Shire, New South Wales. I cheered internally when she told me that she's finally off the antidepressant medication that she's been on for over two decades. The inspiration to do so was planted while sitting the four-day Vision Quest that she and Laura participated in last year. I'll never forget the brightness of the spark of life in her eyes the morning I welcomed her back to the base-camp fire. She told us later in the story council that what had emerged for her in that ceremony was a 'trust in my heart wholeheartedly'.

Instead of using words, Laura emerged from her solo fast singing. 'The songs just came to me,' she said. The experience also ushered in a period of 'unbearable grief that has shaken every cell of my being', as she later wrote in an email — the empty spaces of lockdown affording her time to fully grieve the loss in the same year of both her long-term relationship and her dog. In times of need, Laura leans in even closer to nature as ally and solace. Her yearning for what she really wants is clear and impassioned. 'I want to be on the land, where my bare hands touch the soil, cultivating a deeper connection, and where the calls of the birds are the radio in the morning. I want to welcome over and over a new intimacy with the more-than-human world.' Laura has moved out of her house and is looking for a place to live this dream into reality.

Lockdown for Lucy also meant being locked out of her sit spot in the Royal Botanic Gardens. This, along with an aggressive neighbour's renovations, provided enough of an impetus for her to take a leap into the unknown, taking nothing much more than her 'ugg boots and binoculars' to a cabin on the edge of a lake for some months in order to deepen her search for her inner bilum, her lost inheritance. She tracked back to her days in New Guinea, remembering how she ignored the call of her soul. 'I didn't want to be a doctor. I knew I shouldn't get married. I just wanted to keep walking across the country.' The period of contemplation galvanised Lucy's decision to return only long enough to sell her apartment. 'I'm no longer willing to live a limited life shaped by someone else,' she told me. 'I don't know where to next, but my kids are the only ones worried about that!' Lucy is on the eve of sitting her first Vision Quest, weeks after she has given away most of her belongings. 'Life self-corrects itself somehow, if you listen. Perhaps I had to lose the bilum in order to reach this place again.'

Michelle has nested into motherhood after giving birth to Ollie under the olive tree in her backyard. The decision to homebirth was a powerful way for Michelle to connect with her matrilineal elders and ancestors. 'Being a mum has taught me so much about nature connection, about presence, slowed pace, and finding the adventure and joy in what's right under my feet,' she tells me. During lockdown, Michelle participated in a local council program that had her creating a habitat garden of Indigenous flowering grasses and shrubs in her front yard. 'It gave me a profound sense of belonging to the landscape, a newfound appreciation for the wildness around me, and showed me how I can be an active participant in rewilding our urban areas,' she said.

From our brief communication, Tas has become busier in lockdown, with homeschooling on top of his osteopathic practice. While his sit spot and conversations with birds have 'dropped off the bandwagon', he tells me that he's been exploring more deeply other areas of his existence, both psychological and spiritual. 'While the nature work we did has faded into the background, it will never leave me, and I will pick up the thread again when space opens up. I know I will.'

Day five. Friday. I pull my belt in a notch and hitch up my jeans. It's the last cook-up I'll share with Taj before we go our separate ways on the weekend, and she's ready to impress. On the menu is a feast of acorn-and-nettle fettucine and gnocchi. It takes us three trips up and down the steep hill to transport the equipment, including a giant granite mortar and pestle.

'You guys are crazy,' our friend Tiaan calls as we set out the mobile pantry of preserves and foraged goodies on stumps. I'm definitely losing my sense of rationing now we're in the last days. Tiaan has come to film our final feast and grabs his camera as I get the fire going. Taj begins to mash up pre-cooked potatoes from my garden, along with acorn flour and a big gloop of dark-green pre-blended nettle. The latter swirls through the mix with food-colouring intensity.

I nestle the cast-iron pan on top of two logs, build the fire up underneath, and drop the bunya nuts in. This is another long-awaited culinary moment. Phillip brought them back from Queensland for me, and I've had to be extremely disciplined not to dip into the stash. I can understand why Indigenous gatherings were held in this nut's honour. They even top macadamias for me. I start pounding the nettle, garlic, and sundried tomato in the large granite mortar and pestle, to get it ready for the pesto.

'You two should make a cookbook,' says Tiaan from behind the camera.

'It's not a bad idea,' says Taj. 'We haven't had time to do half the ideas we were planning.'

Tiaan looks up, alerting me to someone coming down the steps. 'Well, here's one for the urban forager's guidebook,' I say dryly, eyeing off the box of gluten-free cookies that Luciano is holding up in one hand.

'No cameras,' says Luciano calls out as he strides towards the fire circle purposefully. I jump up to greet him. 'Shouldn't you be wearing your fox-skin vest for this occasion?' he says.

'It's almost there. Not quite ready for the camera, though,' I smile.

'Escargot for the queen?' he says, extending the box in my direction. It's not cookies inside. Half a dozen long grey antennae are poking out the top. I take a closer look — thirty garden snails, moving about each other.

'Snail man, I presume?' says Taj, coming over to make the acquaintance.

Luciano laughs. 'I see I've been demoted from fox man to snail man.'

'Oh well, if I call you chocolate man will you grow some for me?' I laugh. A snail crawls out into my finger, and I flick it off. I'm surprised by how disgusted I am. But insects and snails are probably the most realistic source of urban protein. Also, Luciano has been feeding them with sourdough bread in a cane basket for the last month in readiness for this moment.

'They're not gluten-free, I can't eat them,' I joke, putting the billy on the fire.

'This used to be my job as a kid,' Luciano tells me. 'Mum would send me out to find the snails, and then they'd boil them up with salt and eat them.'

'So I'll save some for you?' I ask.

'Oh no, I wouldn't take a snail out of the mouth of an urban survivalist.'

I roll my eyes in response.

The water quickly boils, and I cringe as I drop the snails in one by one. Large or small, it's never easy to take a life. After a few minutes, we drain the water and tip them onto a stump. Luciano helps me deshell them.

'This is where oil, salt, and garlic are my best friends,' I say grimly. After the ample addition of all three and the application of high heat, Taj and I stand shoulder to shoulder in solidarity to taste-test our first garden snail. Tiaan crowds in with the camera, eager to catch the moment.

'It's just like calamari,' I say, chewing gingerly. Perhaps what's more true is that anything tastes good when deep-fried with garlic and salt. Taj and I high-five our bravery, then get busy to finish preparing our meal.

I mix up a chickweed salad with pomegranate seeds, bell pepper, and roasted kurrajong seeds, finish crushing the bunya nuts in the pesto, and dress up the rest of the snails in parsley and lemon juice. I'm dishing up the last of the last of the four bowls with blood-red sauerkraut when Taj produces an avocado from her bag of tricks.

'Ta da,' she says proudly. I can't believe it. It's from someone's cousin's girlfriend's neighbour's tree or something. I don't care. It's an avocado. I dreamt last night that I found three large zucchinis hiding in my garden. There's abundance everywhere I turn this week. We slice the avocado in quarters, clink bowls in gratitude, and draw closer to the fire to eat.

I look down into my dinner. The colour palate is astounding. I don't think I've ever beheld a bowl with so much life. I keep my hands cupped around the bowl and continue to stare into it as Taj chats away to Tiaan and Luciano. There's a warmth emanating through the wood of the bowl and into my hands. This is not just a bowl of food. It's a bowl of connections. A bowl of love.

I could pick any ingredient from this bowl, or from any of the bowls I've supped from this week, and share a story of its origin, stories that come from the life that it has lived through my own and my friend's hands. Scaling the roadside cliff to pick wild plums as I watched the tail-lights of my lover's truck disappear around the bend. The four sausage-bent cucumbers that I picked from the home trellis on a hot summer afternoon wearing nothing but a sarong. The conversation I had with Tas while knee-deep in the salt water of the bay collecting Neptune's necklace seaweed. The neighbours I met down the street who allowed me to scale their tree to harvest the corn-yellow kurrajong seeds. Phillip handing me the bag of bunya nuts with a giant smile, his crow's-feet more pronounced as his skin darkened in the northern sun. Elly's daughters singing to the bell peppers. A green juice decorated with peppermint, pepino and tamarillo delivered to me by my man in an outdoor bath. The pungent smell of juniper bark, and the lilt of Meg's sweet voice next to me as I reached in through the spiky shrub to pluck the small hard indigo fruit. The feeling of the late-summer sun on my back as I picked blackberries.

Relating to the monetary economy is straightforward — a linear transaction. This non-monetary economy is far more involved and nuanced. It requires paying attention, the acquisition of knowledge and skill, the slow ripening of community. The number of relationships inherent within this one bowl makes for a more beautiful life, and as such, a more beautiful possibility for culture. If the larger crisis at hand is, as Charles Eisenstein posits, a crisis of belonging, then this bowl is good medicine for it. Food is the basis of belonging, the word 'culture' deriving from the Latin *cultura*, meaning to cultivate the soil. Eating food is the most intimate way a human can interact with the land — soil and flesh becoming flesh. It defines how we interact with the land and each other. This tending to my most basic of needs *belongs* me. It's food born of all my relations — everything from the soil microbes and the bees, to the hands that planted the seed and picked the fruit. It's a sacrament to life itself.

On the final evening, I sit at my beloved's table in his home, nestled between mountains and river. There's a salad grown and gathered by his hands, sprinkled through with calendula petals; the last of the lamb, rubbed with pine needles; native pepper and juniper berries; the beetroot and carrot my parents planted; rhubarb crumble in the oven for dessert; and to drink, elderberry wine gifted to me two years ago by my friend Rosie. It's a wild meal and a meal for the wild. We clink our goblets together and cheers the end of a wonderful week. Tonight, I feast on more than this plate of good food. I feast on my life. To all my relations, I give my deepest thanks.

Acknowledgements

The birth of this book really rests on the back and banks of the river and all its inhabitants — both human and non-human — who accompanied me on this pilgrimage to discover the untamed heart of the urban jungle. What started out as a year experiment turned into a four-plus year odyssey — I didn't want to come inside!

Firstly, gratitude to the Wurundjeri people, the original people of this land I live on, some of whom I have had the pleasure to share a fireside chat with, and to all the Indigenous peoples of the earth, who continue to be keepers of earth's wisdom. Thank you to all the other keepers of nature's knowledge who so generously shared their wisdom and stories with me.

Gratitude to all the patient team at Scribe, particularly my editor, David Golding, who championed the idea from the start and continued to do so throughout the long and tangential journey from conception to birth. Thank you also Laura Thomas for saving the day with the cover design.

To my agent, Gaby Naher, for your unerring support and the way you keep showing up with the finest of cheerleading gestures and words.

I might certainly have faltered were it not for my early readers who offered invaluable encouragement and feedback. To my dear Beth, you saved yet another manuscript from the fire. I can't thank you enough for your insightful and loving attention to my chapters. To David Roland, your feedback was timely and anchoring in the best possible way. To Craig Poulton, for encouraging me to dig deeper into the soul of my story, and for all your love and support along the way. To Arienne Bloodwood, for the backbone of support you offer in continued visible and invisible ways. Gratitude also to Bobbi Allan, Caiyloirch Marques, Sunni Boulton, Min Manifold, Maya Ward, Sarah Allely, Amy Clarkson, Tyson Yunkaporta, and Susie Russell for the gift of intelligent and caring eyes to my words. The writerly nest offered to me at Nest, care of Jay Chubb, was a much needed incubator for this work.

Thanksgiving must go to my mentor and friend Jon Young, firstly for the gift of an incredibly radical and accessible body of work — a true map for how to live a deeply connected life, which I continue to draw from. The maps truly translate onto the territory. And thank you for your unfailing enthusiasm and excitement for the project; for sharing laughs and stories; and for the fierce reminder that creation mind exists in every moment.

Gratitude to Miriam Lancewood for role-modelling how wild women can truly support each other's creative work. And to Richard Louv, who generously picked up the chapters at the last minute with the eyes of one who truly knows what it takes to create a nature-connected village.

Gratitude also to my mentor Bill Plotkin, for putting soul at the heart of my conversation with myself and the world, and that of many others. Your courage to dream the impossible is breathtaking. I join with you in this, the task of our times.

To my wild-hearted bunch of beloved friends, colleagues, and neighbourhood folk of this city that has been my home for the last six-and-a-half years. You colour and inspire both the pages of my book and the pages of my life with joy, creativity, love, and the right amount of tension. Thank you for providing such quirky tales to tell and adventures to be shared. Thanks to the Lighthouse crew for providing a refuge for the writerly heart during lockdown #2.

To all I have and continue to share the home and hearth with in the Riverhouse, especially beloved Megan, who now resides with the ancestors, thank you for being family.

A special note of gratitude to Kate and Cat (aka Babetown) for helping walk home not just this project but the much more diabolical project of Life itself. To Sam, Min, and Chelle for providing a warm refuge for this heart time and time again. I stand taller because of the foundation of strong women in my life.

Thanks to all the Rewild Friday crew for bringing their wild hearts and wild stories and the generosity to share them, and to Mel for being a very easterly conspirator when all my west wanted was to get out of Melbourne's cold!

To the lifelong and untiring support from my parents, Bob and Pauline Dunn. You understandably baulked at the idea of supporting me through another book, and yet backed me nevertheless. To my sister, Liv, for similar sentiments. And to all my beautiful puppy pile of a family, thanks for being crazy cats who I get to roll around with whenever I can cross borders to do so.

A bear hug of appreciation to my beloved Daniel, for your beautiful willingness, your radical acceptance, pressing questions, and wolfish ways. May we continue to track the wild trails together.

And lastly, a heartfelt debt of gratitude to the wild beings of the Birrarung and its creeks and tributaries: the feathered powerful owls, humble ducks, and screeching lorikeets, the slithering tiger snakes and blue-tongue lizards, the furred foxes, possums, and rakalis, the green beings of the giant river red gums and cherry ballarts, and the life-giving waters of the river itself for welcoming me with talon, paw, and claw and for teaching me what I can only learn in the quiet pauses of reverent attention to the seasons, cycles, and mirrors of wild earth. Wild Blessings to all.